Ben Stacy Jerrik (Ed.)

Retail Foreign Exchange Platform

Ben Stacy Jerrik (Ed.)

Retail Foreign Exchange Platform

Foreign exchange market, Spot contract, Electronic trading

Part Press

Publisher:
Part Press is a trademark of
International Book Market Service Ltd., 17 Rue Meldrum, Beau Bassin, 1713-01 Mauritius
Email: info@bookmarketservice.com
Website: www.bookmarketservice.com

Published in 2012

Printed in: U.S.A., U.K., Germany. This book was not produced in Mauritius.

ISBN: 978-620-1-52176-6

Contents

Articles

References

Retail_foreign_exchange_platform

Retail foreign exchange trading is a small segment of the large foreign exchange market. In 2007 it had been speculated that volume from retail foreign exchange trading represents 5 percent of the whole foreign exchange market which amounts to $50–100 billion[1] [2] in daily trading turnover. The retail foreign exchange market has been growing. In general retail customers are able to trade spot currencies. Due to the increasing tendency in the past years of the gradual shift from traditional intrabank 'paper' trading to the more advanced and accurate electronic trading, there has been spur in software development in this field. This change provided different types of trading platforms and tools intended for the use by banks, portfolio managers, retail brokers and retail traders.

One of the most important tools required to perform a foreign exchange transaction is the trading platform providing retail traders and brokers with accurate currency quotes.

History and new developments

Since 1996, when retail foreign exchange trading was first introduced, several brokers who lacked the sufficient tools developed their own trading platforms tailored specifically to their needs. The 1st retail FX brokers were MG Forex, The Matchbook FX ECN, GFT, CMC Markets, Saxo Bank (then known as Midas) and a handful of others. Most except CMC, Saxo & Matchbook FX were based on the ACT foreign exchange trading technology and GUI. These platforms were good enough at the time but required constant investments in research and development and this development cost too much. This was the **first wave**.

The **second wave** was in the early 2000s: several software companies entered the retail foreign exchange trading market by launching their own versions of trading platforms, like Apbg Group, Ctn Systems and MetaTrader 4 from MetaQuotes Software which allowed the users to create their own trading indicators and automatic strategies. Typically these versions were cumbersome for both front-end users (retail traders) and back-end users (retail brokers) due to the misunderstanding of the developers about the foreign exchange market and also because of the insufficient programming tools/languages at the time. Simultaneously most of the retail brokers kept using and developing their own systems as they waited for better platforms which were yet to be developed.

It is only in the last couple of years that the advanced trading platforms started to emerge. Platforms like Multicharts and cTrader put much stronger emphasis on the user interface (GUI) making it more accessible to the retail traders while making trading on it very simple and intuitive. Platforms started to focus on social networking as a way to attract new users, after the emergence of Facebook, Twitter and other social media networks. Social trading has been growing intensely in the last years, especially after platforms like Currensee, Zecco.com, eToro or FXStat appeared.

Moreover, a very strong emphasis was put on the back-end which allowed the retail brokers better control over their operations, better reporting and accurate system and ways to manage marketing campaigns.

Gradually this wave is replacing the previous second wave with a major shift now to the friendlier and more intuitive systems of the **third wave** which according to Aite Group are necessary in order to maintain growth.[3]

Peer-to-peer trading systems

In the South Pacific, "peer-to-peer" foreign exchange services, supported by local government agencies, are emerging in an attempt to reduce transaction costs to heavily remittance-dependent nations, such as Tonga and Samoa.

Banks

In 2006 a few banks have started offering foreign exchange trading services to individual traders and money managers. DBFX and CitiFX Pro are some of the banks that were offering this service. However Deutsche Bank shut down its DBFX service in 2011.

See also

- Broker

References

[1] Triennial Central Bank Survey (http://www.bis.org/publ/rpfxf07t.pdf) (December 2007), Bank for International Settlements
[2] Greenwich Associates Report (http://www.thetradenews.com/trading/foreign-exchange/772)
[3] Aite Group's report about platform retail trading critical issues (http://www.aitegroup.com/reports/200707161.php)

Foreign_exchange_market

The **foreign exchange market** (**forex**, **FX**, or **currency market**) is a form of exchange for the global decentralized trading of international currencies. Financial centers around the world function as anchors of trading between a wide range of different types of buyers and sellers around the clock, with the exception of weekends. The foreign exchange market determines the relative values of different currencies.[1]

The foreign exchange market assists international trade and investment by enabling currency conversion. For example, it permits a business in the United States to import goods from the European Union member states especially Eurozone members and pay Euros, even though its income is in United States dollars. It also supports direct speculation in the value of currencies, and the carry trade, speculation based on the interest rate differential between two currencies.[2]

In a typical foreign exchange transaction, a party purchases a quantity of one currency by paying a quantity of another currency. The modern foreign exchange market began forming during the 1970s after three decades of government restrictions on foreign exchange transactions (the Bretton Woods system of monetary management established the rules for commercial and financial relations among the world's major industrial states after World War II), when countries gradually switched to floating exchange rates from the previous exchange rate regime, which remained fixed as per the Bretton Woods system.

The foreign exchange market is unique because of

- its huge trading volume representing the largest asset class in the world leading to high liquidity;
- its geographical dispersion;
- its continuous operation: 24 hours a day except weekends, i.e. trading from 20:15 GMT on Sunday until 22:00 GMT Friday;
- the variety of factors that affect exchange rates;
- the low margins of relative profit compared with other markets of fixed income; and
- the use of leverage to enhance profit and loss margins and with respect to account size.

As such, it has been referred to as the market closest to the ideal of perfect competition, notwithstanding currency intervention by central banks. According to the Bank for International Settlements,[3] as of April 2010, average daily turnover in global foreign exchange markets is estimated at $3.98 trillion, a growth of approximately 20% over the $3.21 trillion daily volume as of April 2007. Some firms specializing on foreign exchange market had put the average daily turnover in excess of US$4 trillion.[4]

The $3.98 trillion break-down is as follows:

- $1.490 trillion in spot transactions
- $475 billion in outright forwards
- $1.765 trillion in foreign exchange swaps
- $43 billion currency swaps
- $207 billion in options and other products

Market size and liquidity

The foreign exchange market is the most liquid financial market in the world. Traders include large banks, central banks, institutional investors, currency speculators, corporations, governments, other financial institutions, and retail investors. The average daily turnover in the global foreign exchange and related markets is continuously growing. According to the 2010 Triennial Central Bank Survey, coordinated by the Bank for International Settlements, average daily turnover was US$3.98 trillion in April 2010 (vs $1.7 trillion in 1998).[3] Of this $3.98 trillion, $1.5 trillion was spot transactions and $2.5 trillion was traded in outright forwards, swaps and other derivatives.

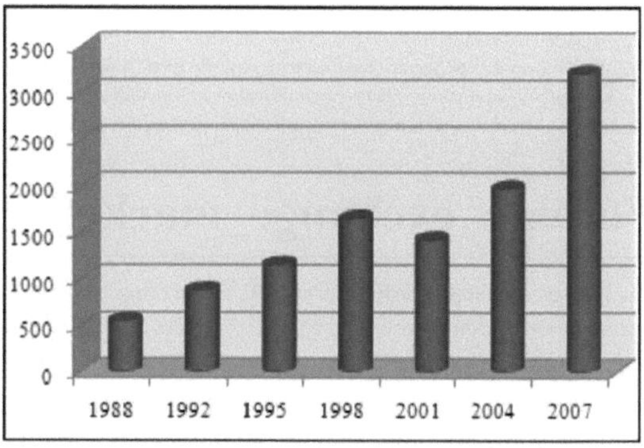

Main foreign exchange market turnover, 1988–2007, measured in billions of USD.

Trading in the United Kingdom accounted for 36.7% of the total, making it by far the most important centre for foreign exchange trading. Trading in the United States accounted for 17.9%, and Japan accounted for 6.2%.[5]

Turnover of exchange-traded foreign exchange futures and options have grown rapidly in recent years, reaching $166 billion in April 2010 (double the turnover recorded in April 2007). Exchange-traded currency derivatives represent 4% of OTC foreign exchange turnover. Foreign exchange futures contracts were introduced in 1972 at the Chicago Mercantile Exchange and are actively traded relative to most other futures contracts.

Most developed countries permit the trading of derivative products (like futures and options on futures) on their exchanges. All these developed countries already have fully convertible capital accounts. Some governments of emerging economies do not allow foreign exchange derivative products on their exchanges because they have capital controls. The use of derivatives is growing in many emerging economies.[6] Countries such as Korea, South Africa, and India have established currency futures exchanges, despite having some capital controls.

Top 10 currency traders [7]
% of overall volume, May 2011

Rank	Name	Market share
1	Deutsche Bank	15.64%
2	Barclays Capital	10.75%
3	UBS AG	10.59%
4	Citi	8.88%
5	JPMorgan	6.43%
6	HSBC	6.26%
7	Royal Bank of Scotland	6.20%
8	Credit Suisse	4.80%
9	Goldman Sachs	4.13%
10	Morgan Stanley	3.64%

Foreign exchange trading increased by 20% between April 2007 and April 2010 and has more than doubled since 2004.[8] The increase in turnover is due to a number of factors: the growing importance of foreign exchange as an asset class, the increased trading activity of high-frequency traders, and the emergence of retail investors as an important market segment. The growth of electronic execution and the diverse selection of execution venues has lowered transaction costs, increased market liquidity, and attracted greater participation from many customer types. In particular, electronic trading via online portals has made it easier for retail traders to trade in the foreign exchange market. By 2010, retail trading is estimated to account for up to 10% of spot turnover, or $150 billion per day (see retail foreign exchange platform).

Foreign exchange is an over-the-counter market where brokers/dealers negotiate directly with one another, so there is no central exchange or clearing house. The biggest geographic trading center is the United Kingdom, primarily London, which according to TheCityUK estimates has increased its share of global turnover in traditional transactions from 34.6% in April 2007 to 36.7% in April 2010. Due to London's dominance in the market, a particular currency's quoted price is usually the London market price. For instance, when the International Monetary Fund calculates the value of its special drawing rights every day, they use the London market prices at noon that day.

Market participants

Unlike a stock market, the foreign exchange market is divided into levels of access. At the top is the interbank market, which is made up of the largest commercial banks and securities dealers. Within the interbank market, spreads, which are the difference between the bid and ask prices, are razor sharp and not known to players outside the inner circle. The difference between the bid and ask prices widens (for example from 0-1 pip to 1-2 pips for a currencies such as the EUR) as you go down the levels of access. This is due to volume. If a trader can guarantee large numbers of transactions for large amounts, they can demand a smaller difference between the bid and ask price, which is referred to as a better spread. The levels of access that make up the foreign exchange market are determined by the size of the "line" (the amount of money with which they are trading). The top-tier interbank market accounts for 53% of all transactions. From there, smaller banks, followed by large multi-national corporations (which need to hedge risk and pay employees in different countries), large hedge funds, and even some of the retail market makers. According to Galati and Melvin, "Pension funds, insurance companies, mutual funds, and other institutional investors have played an increasingly important role in financial markets in general, and in FX markets in particular, since the early 2000s." (2004) In addition, he notes, "Hedge funds have grown markedly over the 2001–2004 period in terms of both number and overall size".[9] Central banks also participate in the foreign exchange market to align currencies to their economic needs.

Commercial companies

An important part of this market comes from the financial activities of companies seeking foreign exchange to pay for goods or services. Commercial companies often trade fairly small amounts compared to those of banks or speculators, and their trades often have little short term impact on market rates. Nevertheless, trade flows are an important factor in the long-term direction of a currency's exchange rate. Some multinational companies can have an unpredictable impact when very large positions are covered due to exposures that are not widely known by other market participants.

Central banks

National central banks play an important role in the foreign exchange markets. They try to control the money supply, inflation, and/or interest rates and often have official or unofficial target rates for their currencies. They can use their often substantial foreign exchange reserves to stabilize the market. Nevertheless, the effectiveness of central bank "stabilizing speculation" is doubtful because central banks do not go bankrupt if they make large losses, like other traders would, and there is no convincing evidence that they do make a profit trading.

Foreign exchange fixing

Foreign exchange fixing is the daily monetary exchange rate fixed by the national bank of each country. The idea is that central banks use the fixing time and exchange rate to evaluate behavior of their currency. Fixing exchange rates reflects the real value of equilibrium in the market. Banks, dealers and traders use fixing rates as a trend indicator.

The mere expectation or rumor of a central bank foreign exchange intervention might be enough to stabilize a currency, but aggressive intervention might be used several times each year in countries with a dirty float currency regime. Central banks do not always achieve their objectives. The combined resources of the market can easily overwhelm any central bank.[10] Several scenarios of this nature were seen in the 1992–93 European Exchange Rate Mechanism collapse, and in more recent times in Southeast Asia.

Hedge funds as speculators

About 70% to 90% of the foreign exchange transactions are speculative. In other words, the person or institution that bought or sold the currency has no plan to actually take delivery of the currency in the end; rather, they were solely speculating on the movement of that particular currency. Hedge funds have gained a reputation for aggressive currency speculation since 1996. They control billions of dollars of equity and may borrow billions more, and thus may overwhelm intervention by central banks to support almost any currency, if the economic fundamentals are in the hedge funds' favor.

Investment management firms

Investment management firms (who typically manage large accounts on behalf of customers such as pension funds and endowments) use the foreign exchange market to facilitate transactions in foreign securities. For example, an investment manager bearing an international equity portfolio needs to purchase and sell several pairs of foreign currencies to pay for foreign securities purchases.

Some investment management firms also have more speculative specialist currency overlay operations, which manage clients' currency exposures with the aim of generating profits as well as limiting risk. While the number of this type of specialist firms is quite small, many have a large value of assets under management), and hence can generate large trades.

Retail foreign exchange traders

Individual Retail speculative traders constitute a growing segment of this market with the advent of retail foreign exchange platforms, both in size and importance. Currently, they participate indirectly through brokers or banks. Retail brokers, while largely controlled and regulated in the USA by the Commodity Futures Trading Commission and National Futures Association have in the past been subjected to periodic Foreign exchange fraud.[11] [12] To deal with the issue, in 2010 the NFA required its members that deal in the Forex markets to register as such (I.e., Forex CTA instead of a CTA). Those NFA members that would traditionally be subject to minimum net capital requirements, FCMs and IBs, are subject to greater minimum net capital requirements if they deal in Forex. A number of the foreign exchange brokers operate from the UK under Financial Services Authority regulations where foreign exchange trading using margin is part of the wider over-the-counter derivatives trading industry that includes Contract for differences and financial spread betting.

There are two main types of retail FX brokers offering the opportunity for speculative currency trading: *brokers* and *dealers* or *market makers*. *Brokers* serve as an agent of the customer in the broader FX market, by seeking the best price in the market for a retail order and dealing on behalf of the retail customer. They charge a commission or mark-up in addition to the price obtained in the market. *Dealers* or *market makers*, by contrast, typically act as principal in the transaction versus the retail customer, and quote a price they are willing to deal at.

Non-bank foreign exchange companies

Non-bank foreign exchange companies offer currency exchange and international payments to private individuals and companies. These are also known as foreign exchange brokers but are distinct in that they do not offer speculative trading but rather currency exchange with payments (i.e., there is usually a physical delivery of currency to a bank account).

It is estimated that in the UK, 14% of currency transfers/payments[13] are made via Foreign Exchange Companies.[14] These companies' selling point is usually that they will offer better exchange rates or cheaper payments than the customer's bank. These companies differ from Money Transfer/Remittance Companies in that they generally offer higher-value services.

Money transfer/remittance companies and bureaux de change

Money transfer companies/remittance companies perform high-volume low-value transfers generally by economic migrants back to their home country. In 2007, the Aite Group estimated that there were $369 billion of remittances (an increase of 8% on the previous year). The four largest markets (India, China, Mexico and the Philippines) receive $95 billion. The largest and best known provider is Western Union with 345,000 agents globally followed by UAE Exchange

Bureaux de change or currency transfer companies provide low value foreign exchange services for travelers. These are typically located at airports and stations or at tourist locations and allow physical notes to be exchanged from one currency to another. They access the foreign exchange markets via banks or non bank foreign exchange companies.

Trading characteristics

Most traded currencies by value

Currency distribution of global foreign exchange market turnover[3]

Rank	Currency	ISO 4217 code (Symbol)	% daily share (April 2010)
1	United States dollar	USD ($)	84.9%
2	Euro	EUR (€)	39.1%
3	Japanese yen	JPY (¥)	19.0%
4	Pound sterling	GBP (£)	12.9%
5	Australian dollar	AUD ($)	7.6%
6	Swiss franc	CHF (Fr)	6.4%
7	Canadian dollar	CAD ($)	5.3%
8	Hong Kong dollar	HKD ($)	2.4%
9	Swedish krona	SEK (kr)	2.2%
10	New Zealand dollar	NZD ($)	1.6%
11	South Korean won	KRW (₩)	1.5%
12	Singapore dollar	SGD ($)	1.4%
13	Norwegian krone	NOK (kr)	1.3%
14	Mexican peso	MXN ($)	1.3%
15	Indian rupee	INR (₹)	0.9%
		Other	12.2%
		Total[15]	**200%**

There is no unified or centrally cleared market for the majority of trades, and there is very little cross-border regulation. Due to the over-the-counter (OTC) nature of currency markets, there are rather a number of interconnected marketplaces, where different currencies instruments are traded. This implies that there is not a *single* exchange rate but rather a number of different rates (prices), depending on what bank or market maker is trading, and where it is. In practice the rates are quite close due to arbitrage. Due to London's dominance in the market, a particular currency's quoted price is usually the London market price. Major trading exchanges include EBS and Reuters, while major banks also offer trading systems. A joint venture of the Chicago Mercantile Exchange and Reuters, called Fxmarketspace opened in 2007 and aspired but failed to the role of a central market clearing mechanism.

The main trading centers are New York and London, though Tokyo, Hong Kong and Singapore are all important centers as well. Banks throughout the world participate. Currency trading happens continuously throughout the day; as the Asian trading session ends, the European session begins, followed by the North American session and then back to the Asian session, excluding weekends.

Fluctuations in exchange rates are usually caused by actual monetary flows as well as by expectations of changes in monetary flows caused by changes in gross domestic product (GDP) growth, inflation (purchasing power parity theory), interest rates (interest rate parity, Domestic Fisher effect, International Fisher effect), budget and trade deficits or surpluses, large cross-border M&A deals and other macroeconomic conditions. Major news is released publicly, often on scheduled dates, so many people have access to the same news at the same time. However, the large banks have an important advantage; they can see their customers' order flow.

Currencies are traded against one another. Each currency pair thus constitutes an individual trading product and is traditionally noted XXXYYY or XXX/YYY, where XXX and YYY are the ISO 4217 international three-letter code of the currencies involved. The first currency (XXX) is the base currency that is quoted relative to the second currency (YYY), called the counter currency (or quote currency). For instance, the quotation *EURUSD (EUR/USD) 1.5465* is the price of the euro expressed in US dollars, meaning 1 euro = 1.5465 dollars. The market convention is to quote most exchange rates against the USD with the US dollar as the base currency (e.g. USDJPY, USDCAD, USDCHF). The exceptions are the British pound (GBP), Australian dollar (AUD), the New Zealand dollar (NZD) and the euro (EUR) where the USD is the counter currency (e.g. GBPUSD, AUDUSD, NZDUSD, EURUSD).

The factors affecting XXX will affect both XXXYYY and XXXZZZ. This causes positive currency correlation between XXXYYY and XXXZZZ.

On the spot market, according to the 2010 Triennial Survey, the most heavily traded bilateral currency pairs were:

- EURUSD: 28%
- USDJPY: 14%
- GBPUSD (also called *cable*): 9%

and the US currency was involved in 84.9% of transactions, followed by the euro (39.1%), the yen (19.0%), and sterling (12.9%) (see table). Volume percentages for all individual currencies should add up to 200%, as each transaction involves two currencies.

Trading in the euro has grown considerably since the currency's creation in January 1999, and how long the foreign exchange market will remain dollar-centered is open to debate. Until recently, trading the euro versus a non-European currency ZZZ would have usually involved two trades: EURUSD and USDZZZ. The exception to this is EURJPY, which is an established traded currency pair in the interbank spot market. As the dollar's value has eroded during 2008, interest in using the euro as reference currency for prices in commodities (such as oil), as well as a larger component of foreign reserves by banks, has increased dramatically. Transactions in the currencies of commodity-producing countries, such as AUD, NZD, CAD, have also increased.

Determinants of exchange rates

The following theories explain the fluctuations in exchange rates in a floating exchange rate regime (In a fixed exchange rate regime, rates are decided by its government):

1. International parity conditions: Relative Purchasing Power Parity, interest rate parity, Domestic Fisher effect, International Fisher effect. Though to some extent the above theories provide logical explanation for the fluctuations in exchange rates, yet these theories falter as they are based on challengeable assumptions [e.g., free flow of goods, services and capital] which seldom hold true in the real world.
2. Balance of payments model (see exchange rate): This model, however, focuses largely on tradable goods and services, ignoring the increasing role of global capital flows. It failed to provide any explanation for continuous appreciation of dollar during 1980s and most part of 1990s in face of soaring US current account deficit.
3. Asset market model (see exchange rate): views currencies as an important asset class for constructing investment portfolios. Assets prices are influenced mostly by people's willingness to hold the existing quantities of assets, which in turn depends on their expectations on the future worth of these assets. The asset market model of exchange rate determination states that "the exchange rate between two currencies represents the price that just balances the relative supplies of, and demand for, assets denominated in those currencies."

None of the models developed so far succeed to explain exchange rates and volatility in the longer time frames. For shorter time frames (less than a few days) algorithms can be devised to predict prices. It is understood from the above models that many macroeconomic factors affect the exchange rates and in the end currency prices are a result of dual forces of demand and supply. The world's currency markets can be viewed as a huge melting pot: in a large and ever-changing mix of current events, supply and demand factors are constantly shifting, and the price of one currency in relation to another shifts accordingly. No other market encompasses (and distills) as much of what is going on in the world at any given time as foreign exchange.

Supply and demand for any given currency, and thus its value, are not influenced by any single element, but rather by several. These elements generally fall into three categories: economic factors, political conditions and market psychology.

Economic factors

These include: (a) economic policy, disseminated by government agencies and central banks, (b) economic conditions, generally revealed through economic reports, and other economic indicators.

- Economic policy comprises government fiscal policy (budget/spending practices) and monetary policy (the means by which a government's central bank influences the supply and "cost" of money, which is reflected by the level of interest rates).
- Government budget deficits or surpluses: The market usually reacts negatively to widening government budget deficits, and positively to narrowing budget deficits. The impact is reflected in the value of a country's currency.
- Balance of trade levels and trends: The trade flow between countries illustrates the demand for goods and services, which in turn indicates demand for a country's currency to conduct trade. Surpluses and deficits in trade of goods and services reflect the competitiveness of a nation's economy. For example, trade deficits may have a negative impact on a nation's currency.
- Inflation levels and trends: Typically a currency will lose value if there is a high level of inflation in the country or if inflation levels are perceived to be rising. This is because inflation erodes purchasing power, thus demand, for that particular currency. However, a currency may sometimes strengthen when inflation rises because of expectations that the central bank will raise short-term interest rates to combat rising inflation.
- Economic growth and health: Reports such as GDP, employment levels, retail sales, capacity utilization and others, detail the levels of a country's economic growth and health. Generally, the more healthy and robust a country's economy, the better its currency will perform, and the more demand for it there will be.
- Productivity of an economy: Increasing productivity in an economy should positively influence the value of its currency. Its effects are more prominent if the increase is in the traded sector [16].

Political conditions

Internal, regional, and international political conditions and events can have a profound effect on currency markets.

All exchange rates are susceptible to political instability and anticipations about the new ruling party. Political upheaval and instability can have a negative impact on a nation's economy. For example, destabilization of coalition governments in Pakistan and Thailand can negatively affect the value of their currencies. Similarly, in a country experiencing financial difficulties, the rise of a political faction that is perceived to be fiscally responsible can have the opposite effect. Also, events in one country in a region may spur positive/negative interest in a neighboring country and, in the process, affect its currency.

Market psychology

Market psychology and trader perceptions influence the foreign exchange market in a variety of ways:

- Flights to quality: Unsettling international events can lead to a "flight to quality", a type of capital flight whereby investors move their assets to a perceived "safe haven". There will be a greater demand, thus a higher price, for currencies perceived as stronger over their relatively weaker counterparts. The U.S. dollar, Swiss franc and gold have been traditional safe havens during times of political or economic uncertainty.[17]
- Long-term trends: Currency markets often move in visible long-term trends. Although currencies do not have an annual growing season like physical commodities, business cycles do make themselves felt. Cycle analysis looks

at longer-term price trends that may rise from economic or political trends.[18]

- "Buy the rumor, sell the fact": This market truism can apply to many currency situations. It is the tendency for the price of a currency to reflect the impact of a particular action before it occurs and, when the anticipated event comes to pass, react in exactly the opposite direction. This may also be referred to as a market being "oversold" or "overbought".[19] To buy the rumor or sell the fact can also be an example of the cognitive bias known as anchoring, when investors focus too much on the relevance of outside events to currency prices.
- Economic numbers: While economic numbers can certainly reflect economic policy, some reports and numbers take on a talisman-like effect: the number itself becomes important to market psychology and may have an immediate impact on short-term market moves. "What to watch" can change over time. In recent years, for example, money supply, employment, trade balance figures and inflation numbers have all taken turns in the spotlight.
- Technical trading considerations: As in other markets, the accumulated price movements in a currency pair such as EUR/USD can form apparent patterns that traders may attempt to use. Many traders study price charts in order to identify such patterns.[20]

Financial instruments

Spot

A spot transaction is a two-day delivery transaction (except in the case of trades between the US Dollar, Canadian Dollar, Turkish Lira, EURO and Russian Ruble, which settle the next business day), as opposed to the futures contracts, which are usually three months. This trade represents a “direct exchange” between two currencies, has the shortest time frame, involves cash rather than a contract; and interest is not included in the agreed-upon transaction.

Forward

One way to deal with the foreign exchange risk is to engage in a forward transaction. In this transaction, money does not actually change hands until some agreed upon future date. A buyer and seller agree on an exchange rate for any date in the future, and the transaction occurs on that date, regardless of what the market rates are then. The duration of the trade can be one day, a few days, months or years. Usually the date is decided by both parties. Then the forward contract is negotiated and agreed upon by both parties.

Swap

The most common type of forward transaction is the swap. In a swap, two parties exchange currencies for a certain length of time and agree to reverse the transaction at a later date. These are not standardized contracts and are not traded through an exchange. A deposit is often required in order to hold the position open until the transaction is completed.

Future

Futures are standardized forward contracts and are usually traded on an exchange created for this purpose. The average contract length is roughly 3 months. Futures contracts are usually inclusive of any interest amounts.

Option

A foreign exchange option (commonly shortened to just FX option) is a derivative where the owner has the right but not the obligation to exchange money denominated in one currency into another currency at a pre-agreed exchange rate on a specified date. The options market is the deepest, largest and most liquid market for options of any kind in the world.

Speculation

Controversy about currency speculators and their effect on currency devaluations and national economies recurs regularly. Nevertheless, economists including Milton Friedman have argued that speculators ultimately are a stabilizing influence on the market and perform the important function of providing a market for hedgers and transferring risk from those people who don't wish to bear it, to those who do.[21] Other economists such as Joseph Stiglitz consider this argument to be based more on politics and a free market philosophy than on economics.[22]

Large hedge funds and other well capitalized "position traders" are the main professional speculators. According to some economists, individual traders could act as "noise traders" and have a more destabilizing role than larger and better informed actors.[23]

Currency speculation is considered a highly suspect activity in many countries. While investment in traditional financial instruments like bonds or stocks often is considered to contribute positively to economic growth by providing capital, currency speculation does not; according to this view, it is simply gambling that often interferes with economic policy. For example, in 1992, currency speculation forced the Central Bank of Sweden to raise interest rates for a few days to 500% per annum, and later to devalue the krona.[24] Former Malaysian Prime Minister Mahathir Mohamad is one well known proponent of this view. He blamed the devaluation of the Malaysian ringgit in 1997 on George Soros and other speculators.

Gregory J. Millman reports on an opposing view, comparing speculators to "vigilantes" who simply help "enforce" international agreements and anticipate the effects of basic economic "laws" in order to profit.[25]

In this view, countries may develop unsustainable financial bubbles or otherwise mishandle their national economies, and foreign exchange speculators made the inevitable collapse happen sooner. A relatively quick collapse might even be preferable to continued economic mishandling, followed by an eventual, larger, collapse. Mahathir Mohamad and other critics of speculation are viewed as trying to deflect the blame from themselves for having caused the unsustainable economic conditions.

Risk aversion

Risk aversion is a kind of trading behavior exhibited by the foreign exchange market when a potentially adverse event happens which may affect market conditions. This behavior is caused when risk averse traders liquidate their positions in risky assets and shift the funds to less risky assets due to uncertainty.[26]

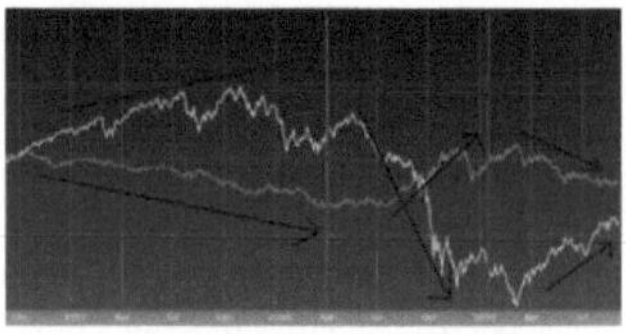

Fig.1 Chart showing MSCI World Index of Equities fell while the US Dollar Index rose.

In the context of the foreign exchange market, traders liquidate their positions in various currencies to take up positions in safe-haven currencies, such as the US Dollar.[27] Sometimes, the choice of a safe haven currency is more of a choice based on prevailing sentiments rather than one of economic statistics. An example would be the Financial Crisis of 2008. The value of equities across the world fell while the US Dollar strengthened (see Fig.1). This happened despite the strong focus of the crisis in the USA.[28]

Carry Trade

Currency carry trade refers to the act of borrowing one currency that has a low interest rate in order to purchase another with a higher interest rate. A large difference in rates can be highly profitable for the trader, especially if high leverage is used. However, with all levered investments this is a double edged sword, and large exchange rate fluctuations can suddenly swing trades into huge losses.

Forex Signals

Forex trade alerts, often referred to as Forex Signals are trade strategies provided by either experienced traders or market analysts. These signals which are often charged a premium fee for can then be copied or replicated by a trader to his own live account. Forex Signal products are packaged as either alerts delivered to a users inbox or sms, or can be installed as an Expert Advisor to a traders MT4 trading platform.

See also

- Balance of trade
- Currency codes
- Currency strength
- Foreign currency mortgage
- Foreign exchange autotrading
- Foreign exchange controls
- Foreign exchange hedge
- Foreign exchange reserves
- Money market
- Nonfarm payrolls
- Special drawing rights
- Tobin tax
- World currency

References

[1] The Economist – Guide to the Financial Markets (https://docs.google.com/fileview?id=0B_Qxj5U7eaJTZTJkODYzN2ItZjE3YyO0Y2M0LTk2ZmUtZGU0NzA3NGI4Y2Y5&hl=en&pli=1) (pdf)

[2] Global imbalances and destabilizing speculation (http://www.igidr.ac.in/~money/mfc_10/Massimiliano_submission_40.pdf) (2007), UNCTAD Trade and development report 2007 (Chapter 1B).

[3] 2010 Triennial Central Bank Survey (http://www.bis.org/publ/rpfxf10t.htm), Bank for International Settlements.

[4] " *What is Foreign Exchange?* (http://au.ibtimes.com/articles/110821/20110210/what-is-foreign-exchange-currency-conversion-financial-markets-forex-foreign-exchange-markets.htm)". Published by the *International Business Times AU* (http://au.ibtimes.com/forex). Retrieved: February 11, 2011.

[5] BIS Triennial Central Bank Survey (http://www.bis.org/publ/rpfx10.pdf), published in September 2010.

[6] "Derivatives in emerging markets" (http://www.bis.org/publ/qtrpdf/r_qt1012f.htm), the Bank for International Settlements, December 13, 2010

[7] Source: Euromoney FX survey FX survey 2011 (http://www.euromoney.com/poll/3301/PollsAndAwards/Foreign-Exchange.html): The Euromoney FX survey is the largest global poll of foreign exchange service providers.'

[8] "The $4 trillion question: what explains FX growth since the 2007 survey? (http://www.bis.org/publ/qtrpdf/r_qt1012e.htm), the Bank for International Settlements, December 13, 2010

[9] Gabriele Galati, Michael Melvin (December 2004). "Why has FX trading surged? Explaining the 2004 triennial survey" (http://www.bis.org/publ/qtrpdf/r_qt0412f.pdf). Bank for International Settlements. .

[10] Alan Greenspan, *The Roots of the Mortgage Crisis: Bubbles cannot be safely defused by monetary policy before the speculative fever breaks on its own.* (http://opinionjournal.com/editorial/feature.html?id=110010981), the Wall Street Journal, December 12, 2007

[11] McKay, Peter A. (2005-07-26). "Scammers Operating on Periphery Of CFTC's Domain Lure Little Guy With Fantastic Promises of Profits" (http://online.wsj.com/article/SB112233850336095645.html?mod=Markets-Main). *The Wall Street Journal* (Dow Jones and Company). . Retrieved 2007-10-31.

[12] Egan, Jack (2005-06-19). "Check the Currency Risk. Then Multiply by 100" (http://www.nytimes.com/2005/06/19/business/yourmoney/19fore.html?_r=2&adxnnl=1&oref=slogin&adxnnlx=1191337503-g1yHfewhqPWye0XtI+Eq0A&oref=slogin). *The New York Times*. . Retrieved 2007-10-30.

[13] The Sunday Times (UK), 16 July 2006

[14] The 5 largest in the UK are Travelex, Moneycorp, HiFX, World First and Currencies Direct

[15] The total sum is 200% because each currency trade always involves a currency pair.

[16] http://papers.ssrn.com/sol3/papers.cfm?abstract_id=711362

[17] Safe haven currency (http://glossary.reuters.com/index.php/Safe_Haven_Currency)

[18] John J. Murphy, *Technical Analysis of the Financial Markets* (New York Institute of Finance, 1999), pp. 343–375.

[19] Investopedia (http://www.investopedia.com/terms/o/overbought.asp)
[20] Sam Y. Cross, *All About the Foreign Exchange Market in the United States* (http://www.newyorkfed.org/education/addpub/usfxm/), Federal Reserve Bank of New York (1998), chapter 11, pp. 113–115.
[21] Michael A. S. Guth, " Profitable Destabilizing Speculation (http://michaelguth.com/economist/chap1.htm)," Chapter 1 in Michael A. S. Guth, Speculative behavior and the operation of competitive markets under uncertainty, Avebury Ashgate Publishing, Aldorshot, England (1994), ISBN 1-85628-985-0.
[22] What I Learned at the World Economic Crisis (http://www.globalpolicy.org/socecon/bwi-wto/critics/2000/whatilearned.htm) Joseph Stiglitz, The New Republic, April 17, 2000, reprinted at GlobalPolicy.org
[23] Summers LH and Summers VP (1989) 'When financial markets work too well: a Cautious case for a securities transaction tax' Journal of financial services
[24] But Don't Rush Out to Buy Kronor: Sweden's 500% Gamble - International Herald Tribune (http://www.iht.com/articles/1992/09/17/perc.php)
[25] Gregory J. Millman, Around the World on a Trillion Dollars a Day, Bantam Press, New York, 1995.
[26] "Risk Averse" (http://www.investopedia.com/terms/r/riskaverse.asp). Investopedia. . Retrieved 2010-02-25.
[27] "Global markets-US stocks rebound, dollar gains on risk aversion" (http://www.reuters.com/article/idUSN0515775320100205). Reuters. 2010-02-05. . Retrieved 2010-02-27.
[28] Stewart, Heather (2008-04-09). "IMF says US crisis is 'largest financial shock since Great Depression'" (http://www.guardian.co.uk/business/2008/apr/09/useconomy.subprimecrisis). London: guardian.co.uk. . Retrieved 2010-02-27.

External links

- A user's guide to the Triennial Central Bank Survey of foreign exchange market activity, Bank for International Settlements (http://www.bis.org/publ/qtrpdf/r_qt1012h.htm)
- London Foreign Exchange Committee with links (on right) to committees in NY, Tokyo, Canada, Australia, HK, Singapore (http://www.bankofengland.co.uk/markets/forex/fxjsc/)
- United States Federal Reserve daily update of exchange rates (http://www.federalreserve.gov/releases/h10/update/)
- Bank of Canada historical (10-year) currency converter and data download (http://www.bankofcanada.ca/en/rates/exchform.html)
- Microstructure effects, bid-ask spreads and volatility in the spot foreign exchange market pre and post-EMU (http://www.cfr.statslab.cam.ac.uk/events/content/20056/mcgroarty2.pdf)
- OECD Exchange rate statistics (monthly averages) (http://stats.oecd.org/Index.aspx?QueryId=169)
- National Futures Association (2010). *Trading in the Retail Off-Exchange Foreign Currency Market* (http://www.nfa.futures.org/nfa-investor-information/publication-library/forex.pdf). Chicago, Illinois.

Spot_contract

In finance, a **spot contract**, **spot transaction**, or simply **spot**, is a contract of buying or selling a commodity, security or currency for settlement (payment and delivery) on the spot date, which is normally two business days after the trade date. The settlement price (or rate) is called **spot price** (or **spot rate**). A spot contract is in contrast with a forward contract or futures contract where contract terms are agreed now but delivery and payment will occur at a future date.

Spot prices and future price expectations

Depending on the item being traded, spot prices can indicate market expectations of future price movements in different ways. For a security or **non-perishable** commodity (e.g. silver), the spot price reflects market expectations of future price movements. In theory, the difference in spot and forward prices should be equal to the finance charges, plus any earnings due to the holder of the security, according to the cost of carry model. For example, on a share the difference in price between the spot and forward is usually accounted for almost entirely by any dividends payable in the period minus the interest payable on the purchase price. Any other cost price would yield an arbitrage opportunity and riskless profit (see rational pricing for the arbitrage mechanics).

In contrast, a **perishable** or soft commodity does not allow this arbitrage – the cost of storage is effectively higher than the expected future price of the commodity. As a result, spot prices will reflect current supply and demand, not future price movements. Spot prices can therefore be quite volatile and move independently from forward prices. According to the unbiased forward hypothesis, the difference between these prices will equal the expected price change of the commodity over the period.

Spot date

To be merged from spot date.

Examples

Bond

Spot rates are estimated via the bootstrapping method, which uses prices of the securities currently trading in market, that is, from the cash or coupon curve. The result is the spot curve, which exists for each of the various classes of securities.

Commodity

A simple example: even if you know tomatoes are cheap in July and will be expensive in January, you can't buy them in July and take delivery in January, since they will spoil before you can take advantage of January's high prices. The July price will reflect tomato supply and demand in July. The forward price for January will reflect the market's expectations of supply and demand in January. July tomatoes are effectively a different commodity from January tomatoes (contrast contango and backwardation).

See also

- Spot market
- Forward contract
- Forward price
- Rational pricing

Electronic_trading

Electronic trading, sometimes called **etrading**, is a method of trading securities (such as stocks, and bonds), foreign exchange or financial derivatives electronically. Information technology is used to bring together buyers and sellers through electronic trading platform and networks to create virtual market places such as NASDAQ, NYSE Arca and Globex which are also known as electronic communications networks (ECNs).

Electronic trading is in contrast to older floor trading and phone trading and has a number of advantages, but glitches and cancelled trades do still occur.[1]

Background

Historically, stock markets were physical locations where buyers and sellers met and negotiated. With the improvement in communications technology in the late 20th century, the need for a physical location became less important, as traders could transact from remote locations.

One of the earliest examples of widespread electronic trading was on Globex, the CME Group's electronic trading platform that allows access to a variety of financial, foreign exchange and commodity markets. The Chicago Board Of Trade produced a rival system that was based on Oak Trading Systems' Oak platform which facilitated 'E Open Outcry,' an electronic trading platform that allowed for electronic trading to take place alongside the trading that took place in the CBOT pits. Oak Trading Systems continues to offer access to global markets via various software applications, including demo packages, and products are available through reputable brokerage firms such as EHedger LLC [2].

Electronic trading makes transactions easier to complete, monitor, clear, and settle. NASDAQ, set up in 1971, was the world's first electronic stock market, though it originally operated as an electronic bulletin board, rather than offering straight-through processing (STP). By early 2007, organizations like the Chicago Mercantile Exchange were creating electronic trading platforms to support the emerging interest in trading within the foreign exchange market.

Today many investment firms on both the buy side and sell side are increasing their spending on technology for electronic trading.[3] Many floor traders and brokers are being removed from the trading process. Traders are relying on algorithms to analyze market conditions and then execute their orders. Trading by humans is largely reserved for block trades, which trades of an unusually large size, in order to limit the market impact of the trade. Increasingly, the trend is to remove human traders not only from the act of trading, but to move trading decisions to an automated basis using Complex Event Processing. [4]

Dates of introduction of electronic trading by the leading exchanges in 120 countries are provided in a Journal of Finance article published in 2005 "Financial market design and the equity premium: Electronic vs. floor trading,".[5]

There are, broadly, two types of trading in the financial markets:

- Business-to-business (B2B) trading, often conducted on exchanges, where large investment banks and brokers trade directly with one another, transacting large amounts of securities, and
- Business-to-consumer (B2C) trading, where retail (e.g. individuals buying and selling relatively small amounts of stocks and shares) and institutional clients (e.g. hedge funds, fund managers or insurance companies, trading far larger amounts of securities) buy and sell from brokers or "dealers", who act as middle-men between the clients

and the B2B markets.

While the majority of retail trading in the United States happens over the Internet, retail trading volumes are dwarfed by institutional, inter-dealer and exchange trading. In developing economies, especially in Asia, retail trading compromises a significant portion of overall trading volume.

Before the advent of electronic trading, exchange trading would typically happen on the floor of an exchange, where traders in brightly colored jackets (to identify which firm they worked for) would shout and gesticulate at one another – a process known as open outcry or "pit trading" (the exchange floors were often pit-shaped – circular, sloping downwards to the centre, so that the traders could see one another).

For instruments which aren't exchange-traded (e.g. U.S. treasury bonds), the inter-dealer market substitutes for the exchange. This is where dealers trade directly with one another or through inter-dealer brokers (i.e. companies like GFI Group, BGC Partners and Garban, who act as middle-men between dealers such as investment banks). This type of trading traditionally took place over the phone but brokers are beginning to offer electronic trading services.

Similarly, B2C trading traditionally happened over the phone and, while much of it still does, more brokers are allowing their clients to place orders using electronic systems. Many retail (or "discount") brokers (e.g. Charles Schwab, E-Trade) went online during the late 1990s and most retail stock-broking probably takes place over the web now.

Larger institutional clients, however, will generally place electronic orders via proprietary electronic trading platforms such as Bloomberg Terminal, Reuters 3000 Xtra, BondsPro, Thomson TradeWeb or CanDeal (which connect institutional clients to several dealers), or using their brokers' proprietary software. For stock trading, the process of connecting counterparties through electronic trading is supported by the Financial Information eXchange (FIX) Protocol. Used by the vast majority of exchanges and traders, the FIX Protocol is the industry standard for pre-trade messaging and trade execution. While the FIX Protocol was developed for trading stocks, it has been further developed to accommodate commodities[6] , foreign exchange[7] , derivatives[8] , and fixed income[9] trading.

Impact

The increase of electronic trading has had some important implications:

- Reduced cost of transactions – By automating as much of the process as possible (often referred to as "straight-through processing" or STP), costs are brought down. The goal is to reduce the incremental cost of trades as close to zero as possible, so that increased trading volumes don't lead to significantly increased costs. This has translated to lower costs for investors.
- Greater liquidity – electronic systems make it easier to allow different companies to trade with one another, no matter where they are located. This leads to greater liquidity (i.e. there are more buyers and sellers) which increases the efficiency of the markets.
- Greater competition – While electronic trading hasn't necessarily lowered the cost of entry to the financial services industry, it has removed barriers within the industry and had a globalisation-style competition effect. For example, a trader can trade futures on Eurex, Globex or LIFFE at the click of a button – he or she doesn't need to go through a broker or pass orders to a trader on the exchange floor.
- Increased transparency – Electronic trading has meant that the markets are less opaque. It's easier to find out the price of securities when that information is flowing around the world electronically.
- Tighter spreads – The "spread" on an instrument is the difference between the best buying and selling prices being quoted; it represents the profit being made by the market makers. The increased liquidity, competition and transparency means that spreads have tightened, especially for commoditised, exchange-traded instruments.

For retail investors, financial services on the web offer great benefits. The primary benefit is the reduced cost of transactions for all concerned as well as the ease and the convenience. Web-driven financial transactions bypass traditional hurdles such as logistics.

Technology and systems

Electronic trading systems are typically proprietary software (*etrading platforms* or electronic trading platforms), running on COTS hardware and operating systems, often using common underlying protocols, such as TCP/IP.

Exchanges typically develop their own systems (sometimes referred to as matching engines), although sometimes an exchange will use another exchange's technology (e.g. e-cbot, the Chicago Board of Trade's electronic trading platform, uses LIFFE's Connect system), and some newer electronic exchanges use 3rd-party specialist software providers (e.g. the Budapest stock exchange and the Moscow Interbank Currency Exchange) use automated trading software originally written and implemented by FMSC, an Australian technology company that was acquired by Computershare, and whose intellectual property rights are now owned by OMX.

Exchanges and ECNs generally offer two methods of accessing their systems –

- an exchange-provided GUI, which the trader runs on his or her desktop and connects directly to the exchange/ECN, and
- an API which allows dealers to plug their own in-house systems directly into the exchange/ECN's.

From an infrastructure point of view, most exchanges will provide "gateways" which sit on a company's network, acting in a manner similar to a proxy, connecting back to the exchange's central system.

ECNs will generally forego the gateway/proxy, and their GUI or the API will connect directly to a central system, across a leased line.

Many brokers develop their own systems, although there are some third-party solutions providers specializing in this area. Like ECNs, brokers will often offer both a GUI and an API (although it's likely that a slightly smaller proportion of brokers offer an API, as compared with ECNs), and connectivity is typically direct to the broker's systems, rather than through a gateway.

Investment banks and other dealers have far more complex technology requirements, as they have to interface with multiple exchanges, brokers and multi-dealer platforms, as well as their own pricing, P&L, trade processing and position-keeping systems. Some banks will develop their own electronic trading systems in-house, but this can be costly, especially when they need to connect to many exchanges, ECNs and brokers. There are a number of companies offering solutions in this area.

Algorithmic trading

Some electronic trades are not planned or executed by human traders, but by complex algorithms.

See also

- eCommerce
- Electronic trading platform
- Stock market data systems
- Straight-through processing (STP)
- Trading room
- 2010 Flash Crash

References

[1] Ian Salisbury and Geoffrey Rogow (September 25, 2008). "Gliches Cancel Electronic Trades" (http://online.wsj.com/article/SB122231165723474365.html). The Wall Street Journal. p. D3.. .
[2] http://www.ehedger.com
[3] ISITC 2011 Member Survey (http://www.isitc.org/pressroom/documents/Member Survey 031111.pdf), March 12 2012.
[4] Trading from the Front: A CEP Hedge Fund (http://fixglobal.com/content/trading-front-cep-hedge-fund), FIXGlobal, December 2011
[5] Jain, Pankaj K., 2005, "Financial market design and the equity premium: Electronic vs. floor trading," Journal of Finance volume 60, issue 6, pp. 2955–2985.
[6] Commodities Trading with FIX (http://fixglobal.com/content/commodities-trading-fix), FIXGlobal, December 2011
[7] FIX in FX (http://fixglobal.com/content/fix-fx) FIXGlobal, December 2010
[8] Global Derivatives - Analysing the key developments and accelerators for FIX (http://fixglobal.com/content/global-derivatives-analysing-key-developments-and-accelerators-fix) FIXGlobal, September 2009
[9] Fixed Income Runs on FIX (http://fixglobal.com/content/fixed-income-runs-fix)

External links

- Electronic Trading Event (http://www.tradetecheurope.com)

Software

Computer software, or just **software**, is a collection of computer programs and related data that provides the instructions for telling a computer what to do and how to do it. Software refers to one or more computer programs and data held in the storage of the computer for some purposes. In other words, software is a set of *programs, procedures, algorithms* and its *documentation* concerned with the operation of a data processing system. Program software performs the function of the program it implements, either by directly providing instructions to the computer hardware or by serving as input to another piece of software. The term was coined to contrast to the old term hardware (meaning physical devices). In contrast to hardware, software "cannot be touched".[1] Software is also sometimes used in a more narrow sense, meaning application software only. Sometimes the term includes data that has not traditionally been associated with computers, such as film, tapes, and records.[2]

Computer software is so called to distinguish it from computer hardware, which encompasses the physical interconnections and devices required to store and execute (or run) the software. At the lowest level, executable code consists of machine language instructions specific to an individual processor. A machine language consists of groups of binary values signifying processor instructions that change the state of the computer from its preceding state. Programs are an ordered sequence of instructions for changing the state of the computer in a particular sequence. It is usually written in high-level programming languages that are easier and more efficient for humans to use (closer to natural language) than machine language. High-level languages are compiled or interpreted into machine language object code. Software may also be written in an assembly language, essentially, a mnemonic representation of a machine language using a natural language alphabet. Assembly language must be assembled into object code via an assembler.

History

The first theory about software was proposed by Alan Turing in his 1935 essay *Computable numbers with an application to the Entscheidungsproblem (Decision problem)*.[3] The term "software" was first used in print by John W. Tukey in 1958.[4] Colloquially, the term is often used to mean application software. In computer science and software engineering, software is all information processed by computer system, programs and data.[4] The academic fields studying software are computer science and software engineering.

The history of computer software is most often traced back to the first software bug in 1946. As more and more programs enter the realm of firmware, and the hardware itself becomes smaller, cheaper and faster as predicted by

Moore's law, elements of computing first considered to be software, join the ranks of hardware. Most hardware companies today have more software programmers on the payroll than hardware designers, since software tools have automated many tasks of Printed circuit board engineers. Just like the Auto industry, the Software industry has grown from a few visionaries operating out of their garage with prototypes. Steve Jobs and Bill Gates were the Henry Ford and Louis Chevrolet of their times, who capitalized on ideas already commonly known before they started in the business. In the case of Software development, this moment is generally agreed to be the publication in the 1980s of the specifications for the IBM Personal Computer published by IBM employee Philip Don Estridge. Today his move would be seen as a type of crowd-sourcing.

Until that time, software was *bundled* with the hardware by Original equipment manufacturers (OEMs) such as Data General, Digital Equipment and IBM. When a customer bought a minicomputer, at that time the smallest computer on the market, the computer did not come with Pre-installed software, but needed to be installed by engineers employed by the OEM. Computer hardware companies not only bundled their software, they also placed demands on the location of the hardware in a refrigerated space called a computer room. Most companies had their software on the books for 0 dollars, unable to claim it as an asset (this is similar to financing of popular music in those days). When Data General introduced the Data General Nova, a company called Digidyne wanted to use its RDOS operating system on its own hardware clone. Data General refused to license their software (which was hard to do, since it was on the books as a free asset), and claimed their "bundling rights". The Supreme Court set a precedent called Digidyne v. Data General in 1985. The Supreme Court let a 9th circuit decision stand, and Data General was eventually forced into licensing the Operating System software because it was ruled that restricting the license to only DG hardware was an illegal *tying arrangement*.[5] Soon after, IBM 'published' its DOS source for free, and Microsoft was born. Unable to sustain the loss from lawyer's fees, Data General ended up being taken over by EMC Corporation. The Supreme Court decision made it possible to value software, and also purchase Software patents. The move by IBM was almost a protest at the time. Few in the industry believed that anyone would profit from it other than IBM (through free publicity). Microsoft and Apple were able to thus cash in on 'soft' products. It is hard to imagine today that people once felt that software was worthless without a machine. There are many successful companies today that sell only software products, though there are still many common software licensing problems due to the complexity of designs and poor documentation, leading to patent trolls.

With open software specifications and the possibility of software licensing, new opportunities arose for software tools that then became the de facto standard, such as DOS for operating systems, but also various proprietary word processing and spreadsheet programs. In a similar growth pattern, proprietary development methods became standard Software development methodology.

Types of software

Software includes all the various forms and roles that digitally stored *data* may have and play in a computer (or similar system), regardless of whether the data is used as *code* for a CPU, or other interpreter, or whether it represents other kinds of information. Software thus encompasses a wide array of products that may be developed using different techniques such as ordinary programming languages, scripting languages, microcode, or an FPGA configuration.

The types of software include web pages developed in languages and frameworks like HTML, PHP, Perl, JSP, ASP.NET, XML, and desktop applications like OpenOffice.org, Microsoft Word developed in languages like C, C++, Objective-C, Java, C#, or Smalltalk. Application software usually runs on an underlying software operating systems such as Linux or Microsoft Windows. Software (or firmware) is also used in video games and for the configurable parts of the logic systems of automobiles, televisions, and other consumer electronics.

Practical computer systems divide software systems into three major classes: system software, programming software and application software, although the distinction is arbitrary, and often blurred.

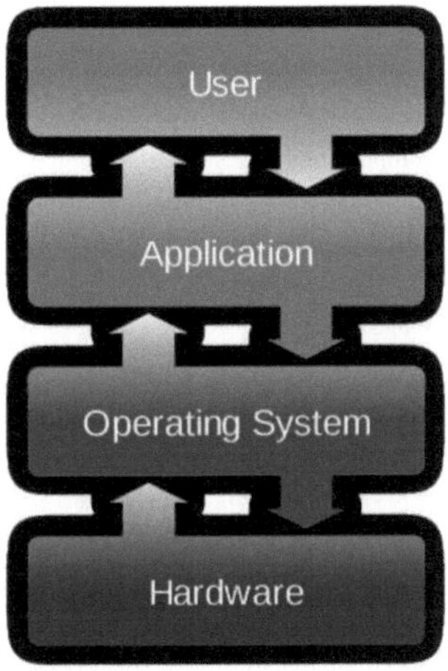

A layer structure showing where the operating system software and application software are situated while running on a typical desktop computer

System software

System software is computer software designed to operate the computer hardware to provide basic functionality and to provide a platform for running application software.[6] [7] System software includes device drivers, operating systems, servers, utilities, and window systems.

System software is responsible for managing a variety of independent hardware components, so that they can work together harmoniously. Its purpose is to unburden the application software programmer from the often complex details of the particular computer being used, including such accessories as communications devices, printers, device readers, displays and keyboards, and also to partition the computer's resources such as memory and processor time in a safe and stable manner.

Programming software

Programming software include tools in the form of programs or applications that software developers use to create, debug, maintain, or otherwise support other programs and applications. The term usually refers to relatively simple programs such as compilers, debuggers, interpreters, linkers, and text editors, that can be combined together to accomplish a task, much as one might use multiple hand tools to fix a physical object. Programming tools are intended to assist a programmer in writing computer programs, and they may be combined in an integrated development environment (IDE) to more easily manage all of these functions.

Application software

Application software is developed to perform in any task that benefits from computation. It is a set of programs that allows the computer to perform a specific data processing job for the user. It is a broad category, and encompasses software of many kinds, including the internet browser being used to display this page. This category includes:

- Business software
- Computer-aided design
- Databases
- Decision-making software
- Educational software
- Image editing
- Industrial automation
- Mathematical software
- Medical software
- Molecular modeling software
- Quantum chemistry and solid state physics software
- Simulation software
- Spreadsheets
- Telecommunications (i.e., the Internet and everything that flows on it)
- Video editing software
- Video games
- Word processing

Software topics

Architecture

Users often see things differently than programmers. People who use modern general purpose computers (as opposed to embedded systems, analog computers and supercomputers) usually see three layers of software performing a variety of tasks: platform, application, and user software.

- Platform software: Platform includes the firmware, device drivers, an operating system, and typically a graphical user interface which, in total, allow a user to interact with the computer and its peripherals (associated equipment). Platform software often comes bundled with the computer. On a PC you will usually have the ability to change the platform software.
- Application software: Application software or Applications are what most people think of when they think of software. Typical examples include office suites and video games. Application software is often purchased separately from computer hardware. Sometimes applications are bundled with the computer, but that does not change the fact that they run as independent applications. Applications are usually independent programs from the operating system, though they are often tailored for specific platforms. Most users think of compilers, databases, and other "system software" as applications.
- User-written software: End-user development tailors systems to meet users' specific needs. User software include spreadsheet templates and word processor templates. Even email filters are a kind of user software. Users create this software themselves and often overlook how important it is. Depending on how competently the user-written software has been integrated into default application packages, many users may not be aware of the distinction between the original packages, and what has been added by co-workers.

Documentation

Most software has software documentation so that the end user can understand the program, what it does, and how to use it. Without clear documentation, software can be hard to use—especially if it is very specialized and relatively complex like Photoshop or AutoCAD.

Developer documentation may also exist, either with the code as comments and/or as separate files, detailing how the programs works and can be modified.

Library

An executable is almost always not sufficiently complete for direct execution. Software libraries include collections of functions and functionality that may be embedded in other applications. Operating systems include many standard Software libraries, and applications are often distributed with their own libraries.

Standard

Since software can be designed using many different programming languages and in many different operating systems and operating environments, software standard is needed so that different software can understand and exchange information between each other. For instance, an email sent from a Microsoft Outlook should be readable from Yahoo! Mail and vice versa.

Execution

Computer software has to be "loaded" into the computer's storage (such as the hard drive or memory). Once the software has loaded, the computer is able to *execute* the software. This involves passing instructions from the application software, through the system software, to the hardware which ultimately receives the instruction as machine code. Each instruction causes the computer to carry out an operation – moving data, carrying out a computation, or altering the control flow of instructions.

Data movement is typically from one place in memory to another. Sometimes it involves moving data between memory and registers which enable high-speed data access in the CPU. Moving data, especially large amounts of it, can be costly. So, this is sometimes avoided by using "pointers" to data instead. Computations include simple operations such as incrementing the value of a variable data element. More complex computations may involve many operations and data elements together.

Quality and reliability

Software quality is very important, especially for commercial and system software like Microsoft Office, Microsoft Windows and Linux. If software is faulty (buggy), it can delete a person's work, crash the computer and do other unexpected things. Faults and errors are called "bugs." Many bugs are discovered and eliminated (debugged) through software testing. However, software testing rarely – if ever – eliminates every bug; some programmers say that "every program has at least one more bug" (Lubarsky's Law). All major software companies, such as Microsoft, Novell and Sun Microsystems, have their own software testing departments with the specific goal of just testing. Software can be tested through unit testing, regression testing and other methods, which are done manually, or most commonly, automatically, since the amount of code to be tested can be quite large. For instance, NASA has extremely rigorous software testing procedures for many operating systems and communication functions. Many NASA based operations interact and identify each other through command programs called software. This enables many people who work at NASA to check and evaluate functional systems overall. Programs containing command software enable hardware engineering and system operations to function much easier together.

License

The software's license gives the user the right to use the software in the licensed environment. Some software comes with the license when purchased off the shelf, or an OEM license when bundled with hardware. Other software comes with a free software license, granting the recipient the rights to modify and redistribute the software. Software can also be in the form of freeware or shareware.

Patents

Software can be patented in some but not all countries; however, software patents can be controversial in the software industry with many people holding different views about it. The controversy over software patents is about specific algorithms or techniques that the software contains, which may not be duplicated by others and considered intellectual property and copyright infringement depending on the severity.

Design and implementation

Design and implementation of software varies depending on the complexity of the software. For instance, design and creation of Microsoft Word software will take much more time than designing and developing Microsoft Notepad because of the difference in functionalities in each one.

Software is usually designed and created (coded/written/programmed) in integrated development environments (IDE) like Eclipse, Emacs and Microsoft Visual Studio that can simplify the process and compile the program. As noted in different section, software is usually created on top of existing software and the application programming interface (API) that the underlying software provides like GTK+, JavaBeans or Swing. Libraries (APIs) are categorized for different purposes. For instance, JavaBeans library is used for designing enterprise applications, Windows Forms library is used for designing graphical user interface (GUI) applications like Microsoft Word, and Windows Communication Foundation is used for designing web services. Underlying computer programming concepts like quicksort, hashtable, array, and binary tree can be useful to creating software. When a program is designed, it relies on the API. For instance, if a user is designing a Microsoft Windows desktop application, he/she might use the .NET Windows Forms library to design the desktop application and call its APIs like *Form1.Close()* and *Form1.Show()*[8] to close or open the application and write the additional operations him/herself that it need to have. Without these APIs, the programmer needs to write these APIs him/herself. Companies like Sun Microsystems, Novell, and Microsoft provide their own APIs so that many applications are written using their software libraries that usually have numerous APIs in them.

Computer software has special economic characteristics that make its design, creation, and distribution different from most other economic goods.[9] [10]

A person who creates software is called a programmer, software engineer, software developer, or code monkey, terms that all have a similar meaning.

Industry and organizations

A great variety of software companies and programmers in the world comprise a software industry. Software can be quite a profitable industry: Bill Gates, the founder of Microsoft was the richest person in the world in 2009 largely by selling the Microsoft Windows and Microsoft Office software products. The same goes for Larry Ellison, largely through his Oracle database software. Through time the software industry has become increasingly specialized.

Non-profit software organizations include the Free Software Foundation, GNU Project and Mozilla Foundation. Software standard organizations like the W3C, IETF develop software standards so that most software can interoperate through standards such as XML, HTML, HTTP or FTP.

Other well-known large software companies include Novell, SAP, Symantec, Adobe Systems, and Corel, while small companies often provide innovation.

See also

- List of software
- Personal computer hardware

References

[1] "Wordreference.com: WordNet 2.0" (http://www.wordreference.com/definition/software). Princeton University, Princeton, NJ. . Retrieved 2007-08-19.

[2] "software..(n.d.)." (http://dictionary.reference.com/browse/software). *Dictionary.com Unabridged (v 1.1)*. . Retrieved 2007-04-13.

[3] Hally, Mike (2005). *Electronic brains/Stories from the dawn of the computer age*. London: British Broadcasting Corporation and Granta Books. p. 79. ISBN 1-86207-663-4.

[4] "John Tukey, 85, Statistician; Coined the Word 'Software'" (http://query.nytimes.com/gst/fullpage.html?res=9500E4DA173DF93BA15754C0A9669C8B63&scp=1&sq=&pagewanted=1). *Obituaries* (New York Times). July 28, 2000. .

[5] "Tying Arrangements and the Computer Industry: Digidyne Corp. vs. Data General". JSTOR 1372482.

[6] (http://depts.alverno.edu/cil/mod1/software/system.html)

[7] "What is software? - Definition from Whatis.com" (http://searchsoa.techtarget.com/sDefinition/0,,sid26_gci213024,00.html). Searchsoa.techtarget.com. 2012-05-13. . Retrieved 2012-05-18.

[8] "MSDN Library" (http://msdn.microsoft.com/en-us/library/default.aspx). . Retrieved 2010-06-14.

[9] v. Engelhardt, Sebastian (2008). "The Economic Properties of Software" (http://ideas.repec.org/p/jrp/jrpwrp/2008-045.html). *Jena Economic Research Papers* **2** (2008–045.). .

[10] Kaminsky, Dan (1999). "Why Open Source Is The Optimum Economic Paradigm for Software" (http://dankaminsky.com/1999/03/02/69/). .

External links

- Software Wikia

Electronic_trading_platform

In finance, an **Electronic trading platform** is a computer system that can be used to place orders for financial products over a network with a financial intermediary. This includes products such as shares, bonds, currencies, commodities and derivatives with a financial intermediary, such as a brokers, market makers, Investment banks or stock exchanges. Such platforms allow electronic trading to be carried out by users from any location and are in contrast to traditional floor trading using open outcry and telephone based trading.

an Electronic Trading Platform being used at the Deutsche Börse

Electronic trading platforms typically stream live market prices on which users can trade and may provide additional trading tools, such as charting packages, news feeds and account management functions. Some platforms have been specifically designed to allow individuals to gain access to financial markets that could traditionally only be accessed by specialist trading firms such as allowing margin trading on forex and derivatives such as contract for difference. They may also be designed to automatically trade specific strategies based on technical analysis or to do high-frequency trading.

Etymology

The term 'trading platform' is generally used to avoid confusion with 'trading system' which is more often associated with the trading method or strategy rather than the computer system used to execute orders within financial circles.[1]

Historic development

The first electronic trading platforms were typically associated with stock exchanges and allowed brokers to place orders remotely using private dedicated networks and dumb terminals. Early systems would not always provide live streaming prices and instead allowed brokers or clients to place an order which would be confirmed some time later, these were known as 'request for quote' based systems.

Trading systems evolved to allow for live streaming prices and near instant execution of orders as well as using the internet as the underlying network meaning that location became much less relevant. Some electronic trading platforms have built in scripting tools and even APIs allowing traders to develop automatic or algorithmic trading systems and robots, which have been used by high frequency traders.

The client graphical user interface of the electronic trading platforms can be used to trade currencies, equities, future, or options and are also sometimes called trading turrets (this may be true, but is probably a misues of the term, as trading turret refers to the specialized phone appliance used by traders).

Trading platform standards

The National Futures Association (NFA) in the US lists the following general standard requirements for forex electronic trading systems;[2]

- Authentication (Passwords, Authentication Tokens such as secureID cards; or digital certificates)
- Encryption
- Transactions recording standard
- Pricing and slippage standard

See also

- Electronic communication network (ECN)
- Retail forex platform
- Stock market data systems
- Straight-through processing (STP)
- Technical analysis software
- Trading system
- Trading room
- Automated trading system

References

[1] "Trading Platforms" (http://www.ibsintelligence.com/index.php?option=com_labels&view=label&label=tradingplatforms). IBS Intelligence. . Retrieved 10 June 2010.
[2] NFA (http://www.nfa.futures.org/printerFriendly.asp?tag=60) (December 2007), NFA Manual

CMC_Markets

CMC cmc markets	
Type	Private (Peter Cruddas and family 86%, Goldman Sachs 10%)
Industry	Financial Services
Founded	London, UK (1986)
Founder(s)	Peter Cruddas
Headquarters	London, UK
Key people	Peter Cruddas(Founder/Chairman), Doug Richards CEO
Services	On-line trading, CFDs and Spread Betting
Revenue	£193.2m (2009), ▲ 7% from 2008
Operating income	£25.3m (2009), ▼ 57% from 2008
Net income	£-15.4m (2009), ▼ 138% from 2008
Total assets	£89.7m (2009), ▼ 13% from 2008
Employees	678 (2011), ▼ was 720 in 2010
Website	www.cmcmarkets.com [1]

CMC Markets is a UK based financial derivatives dealer. The company offers online trading in spread betting, Contract for difference (CFDs) and Foreign Exchange (Forex) across world markets. While large part of its activity is in the UK, Europe and Australia, it has presence internationally.

History

The company was founded in 1989 by Peter Cruddas as a Foreign Exchange market maker under the name 'Currency Management Corporation'. The name was later abbreviated to CMC and then changed to CMC Markets in September 2005.

CMC Markets also operated the brand name deal4free.com from 2001 to 2005. The deal4free brand was designed to promote the zero commission charge service which it used primarily for its spread betting business in the UK. Commissions were later re-introduced and this brand was a dropped as part of the re-branding in September 2005.[2]

In 1992, CMC Markets became authorised and regulated in the UK by the AFBD which later became the Financial Services Authority (FSA)

In 1996, the company launched a real-time FX trading platform and has made claims to have done the first FX deal on-line over the internet. It is hard to verify, but even if it was not true, CMC Markets was certainly one of the first companies to offer on-line trading over the internet. As well as being a pioneer in internet trading technology with its MarketMaker software platform.

The software was developed by Information Internet Limited a sister company jointly owned by Peter Cruddas and the two developers that designed and built the first version of the software, Terry Johnston and Ben Fisher. From 1996 to 2000 Information Internet as well as supplying CMC Markets, sold the software to a small number of banks. In 2000 Peter Cruddas bought out the other owners, made the software exclusive for CMC Markets and the remaining parts of Information Internet became the internal IT function of CMC Markets.

In 2000, CMC Markets began to offer Contracts for difference (CFDs) and the following year it introduced on-line Spread Betting on financial markets. These two products now make up the bulk of CMC Markets business.

In 2002, CMC Markets opened its first Asian office in Sydney, Australia followed in 2005 by Beijing and Hong Kong.

The first European office was opened in Frankfurt in 2005, followed by Stockholm in 2007 and Madrid in 2008.

The first North American office was opened in 2003 in New York this was followed by the acquisition of Canadian broker Shorecan Index which became its Toronto office in 2005. CMC Markets subsequently pulled out of the US market and closed its New York office in 2008. Primarily as it was unable to offer its main CFD and spread betting products there due to regulatory restrictions.

CMC Markets expanded quickly from 2002 to 2007 and opened offices in a large number of countries as well as growing the spread betting business in the UK and CFDs internationally. It offered services in a number of different languages and was one of the first CFD provider in Australia to offer full Chinese-speaking service. A service they expanded globally.

In 2006 Peter Cruddas planned to float the company and the company had been prepared for IPO.[3] [4] However the IPO was cancelled at the last minute with Peter Cruddas citing market conditions.

In 2007 it bought financial media and technology company 'digital look',[5] which ran the financial information site www.digitallook.com [6] as well as selling financial data to third party sites. The company was merged into CMC markets main operations in London, but continues to run the web site and data services for third parties. In 2008 CMC Markets bought the Australian stock broker 'Andrew West', this was merged with the CMC Markets Australian operation and renamed CMC Markets StockBroking. This continues to offer physical share broking services in Australia.

In 2007 Goldman Sachs bought a 10% stake in the company[7] and the due diligence that it did as part of the purchase resulted in a overhaul of the company direction.

In 2008 and 2009 CMC Markets saw profits decline with GFC, in response Peter Cruddas changed his management team, let 350 staff go and closed seven offices.[8]

In 2010, CMC Markets launched its new trading platform[9] and iPhone application[10] to the UK market. The new trading software featured a new pricing engine with 100% automated execution, prices quoted to additional decimal points and no re-quotes.

In 2011 the Digital Look business unit was sold to the Spanish based Web Financial Group.[11]

On 24 March 2012 the owner of CMC markets was secretly videotaped by the Sunday Times allegedly selling access to the UK Prime Minister for £250,000.00 http://www.bbc.co.uk/news/uk-politics-17501618

Operations

CMC Markets operates and has offices in the UK, Australia, Germany, New Zealand, Singapore, Spain, Ireland, Italy, Japan, China, Sweden, Norway, Denmark, Austria, France and Canada.

CMC Markets runs most of its operations from its head office in London and runs a trading desk in London and in Singapore.

Competition

CMC Markets is one of the companies that launched CFDs products to the retail market in 2000 and is one of the largest CFD providers globally. Actual market share is difficult to confirm as there is no reliable independent data in the over-the-counter CFD and Spread Betting industry as providers have no duty to report trading volumes or transactions. In 2009 some industry participants commissioned a survey to look at the UK market and results indicated that CMC Markets market share in 2008 was lower than its main competitor.[12]

References

[1] http://www.cmcmarkets.com/

[2] CMC Group announces rebranding (http://www.encyclopedia.com/doc/1G1-136265969.html)

[3] Volatile market conditions force CMC to pull float (http://www.independent.co.uk/news/business/news/volatile-market-conditions-force-cmc-to-pull-float-479661.html)

[4] Peter Cruddas prepares for CMC flotation (http://www.telegraph.co.uk/finance/newsbysector/banksandfinance/6024184/Peter-Cruddas-prepares-for-CMC-flotation.html)

[5] Cruddas appointments bolster CMC board (http://www.telegraph.co.uk/finance/markets/2784657/Cruddas-appointments-bolster-CMC-board.html)

[6] http://www.digitallook.com/

[7] Goldmans buys 10% stake in CMC Markets (http://business.timesonline.co.uk/tol/business/industry_sectors/banking_and_finance/article2921840.ece)

[8] Peter Cruddas has taken CMC Markets back to basics (http://www.telegraph.co.uk/finance/newsbysector/banksandfinance/6030980/Peter-Cruddas-has-taken-CMC-Markets-back-to-basics.html)

[9] (http://www.financial-spread-betting.com/news/annoucements/26741-cmc-markets-revamps-trading-platform)

[10] (http://stockmarketsreview.com/pressrelease/2010/09/11/cmc-markets-launches-new-spread-betting-app-for-the-iphone/)

[11] Web Financial Group Press Release (http://corporate.webfg.com/en/webfgDl_20111026.htm)

[12] Investors switch between spread betting providers (http://www.ft.com/cms/s/2/b42f3678-ce23-11de-a1ea-00144feabdc0.html)

External links

- CMC Markets plc (http://www.cmcmarketsplc.com/)
- CMC Markets plc (http://www.cmcmarkets.com/)
- CMC Markets UK (http://www.cmcmarkets.co.uk/)
- CMC Markets Singapore (http://www.cmcmarkets.com.sg/)

Portfolio_manager

A **portfolio manager** is either a person who makes investment decisions using money other people have placed under his or her control or a person who manages a financial institution's asset and liability (loan and deposit) portfolios.

On the investments side, they work with a team of analysts and researchers, and are ultimately responsible for establishing an investment strategy, selecting appropriate investments and allocating each investment properly for a fund- or asset-management vehicle.

Portfolio managers are presented with investment ideas from internal buy-side analysts and sell-side analysts from investment banks. It is their job to sift through the relevant information and use their judgment to buy and sell securities. Throughout each day, they read reports, talk to company managers and monitor industry and economic trends looking for the right company and time to invest the portfolio's capital.

A team of analysts and researchers are ultimately responsible for establishing an investment strategy, selecting appropriate investments and allocating each investment properly for a fund or asset-management vehicle.

Portfolio managers make decisions about investment mix and policy, matching investments to objectives, asset allocation for individuals and institutions, and balancing risk against performance.

Portfolio management is about strengths, weaknesses, opportunities and threats in the choice of debt vs. equity, domestic vs. international, growth vs. safety, and other tradeoffs encountered in the attempt to maximize return at a given appetite for risk.

In the case of mutual and exchange-traded funds (ETFs), there are two forms of portfolio management: passive and active. Passive management simply tracks a market index, commonly referred to as indexing or index investing. Active management involves a single manager, co-managers, or a team of managers who attempt to beat the market return by actively managing a fund's portfolio through investment decisions based on research and decisions on individual holdings. Closed-end funds are generally actively managed..

Saxo_Bank

SAXO BANK	
Type	Private company
Founded	1992
Headquarters	Copenhagen, Denmark
Key people	Kurt K. Larsen, Chairman Dennis Malamatinas, Vice-Chairman Kim Fournais, Chief Executive Officer Lars Seier Christensen, Chief Executive Officer
Products	Investment Banking
Revenue	▲ DKK 3,338.1 Million (2010)
Net income	▲ DKK 643.5 Million (2010)
Employees	964 (Dec. 2010)
Website	www.saxobank.com [1]

Saxo Bank is an online Danish investment bank. It was founded as a brokerage firm[2] in 1992, under the name Midas Fondsmæglerselskab, by Lars Seier Christensen, Kim Fournais and Marc Hauschildt. The name was changed to Saxo when the company obtained a banking license in 2001. Saxo offers trading through its online platform SaxoTrader in Forex, Stocks, CFDs, Futures, Funds and Bonds. Private Wealth Management services are also offered. According to Saxo, roughly two-thirds of its activities are derived from partnerships with institutional trading partners. The company functions as an online broker with a bank license, without offering traditional banking products.

Internationally, Saxo Bank is known for its success in Internet brokerage and has received a number of awards. Domestically the bank is also known for its two founders who are often outspoken on Danish politics. Following a record income and profit in 2008, the bank's net profit ended at DKK 201 million in 2009 with 80% earned in the second half of the year.[3] [4]

In August 2010, the bank's half-year results revealed a net profit of 551 million DKK, which is more than any other full year results in Saxo Bank's history.[5] Following the half-year results, rumours emerged that part of the bank is up for sale.[6] Following its record 2010 results, Saxo Bank holds a market value of about 20 billion DKK, according to media reports.[7]

Since late 2010, Saxo Bank has launched several joint White Label solutions with industry giants such as Microsoft, Barclays Stockbrokers Ltd., the financial services company, and TD Waterhouse, UK's leading execution-only brokers.[8] [9] [10]

History

1990s: Midas Fondsmæglerselskab

Lars Seier Christensen and Kim Fournais formed Midas Fondsmæglerselskab after meeting in London in the early 1990s. Christensen was working as a broker for Gerald Metals at the time while Fournais worked at the now defunct Lannung Bank. The two said that they both felt like they had found their business soul mate.[11]

Midas initially operated as a broker selling option products without obtaining the required licenses. When the company acquired its license, the periodical *Økonomisk Ugebrev*, *Economic Weekly* began publishing an article series on *sidegadevekslererne*, the bucket shops of Denmark, and included Midas.[12] [13] [14]

During the autumn of 1996 well-known Danish business man Karsten Ree claimed he had been cheated by Midas into making fictive investments via their wealth management services, suffering losses in excess of 800.000 DKK in one month.[15] The company subsequently underwent a criminal investigation by the Director of Public Prosecutions but was cleared of all charges[12] .

2000s: Name change

In 1997 Midas launched its first internet product; a trading platform for currencies called *MITS* (later replaced by *SaxoTrader*). The company renamed to Saxo Bank, named after the Danish Saxo Grammaticus, because there was already a Nigerian bank called Mida operating at an international level. Saxo obtained its new banking license in 2001, began expanding its online presence and moved to new headquarters in the Copenhagen suburb of Gentofte.

Saxo Bank also decided to outsource work on their new software to a Russian company, Reksoft, which worked on the software from August 2003 to December 2006.[16]

Global expansion

Saxo Bank launched its products and services in Europe during the early 2000s. In 2001, its first White Label solution was launched with a Portuguese securities dealer.[17] In September 2004, it also launched its first US white label solution. Following the launch of Citigroup's CitiFX Pro, an online foreign exchange trading platform developed by Saxo, in November 2008[18] , the Bank would focus entirely on its institutional clients in the US.[19]

In March 2004, Saxo Bank had to halt its business in Australia[20] because the Australian Securities and Investments Commission (ASIC) passed regulation requiring an Australian bank license to offer foreign exchange to private customers[20] . Saxo had to terminate all its Australian retail clients and passed the accounts to two Australian institutional clients[20] . Saxo had particular problems with the Australian authorities, as Danish legislation did not allow the country's regulator to sign a bilateral agreement with them[20] . On 17 September, the issue was resolved, and ASIC granted Saxo a license to trade in Australia.[20] Since then Saxo has also mainly focussed on institutional solutions in the Oceanian countries and territories.

In 2005, it won FXweek's e-FX Award, in the category *Best Retail Platform* and reported a 39% increase in pre-tax profits, while net profits rose DKK 161 million (EUR 21.6 million) to DKK 223 million. In 2006, Saxo expanding its presence primarily through white label clients and made changes to its structure, seeking to boost its retail forex with hires from the Danish business world. Internal restructuring took place and Lars Seier Christensen said the changes were made to reflect growth.[21]

Acquisitions

In September 2007, Saxo announced the acquisition of its Swiss white label client, Synthesis Bank. In late May, Saxo bought a French brokerage, Cambiste, that was later renamed Saxo Banque France. During 2009, Saxo opened offices in Milan, Prague and Dubai. By year-end, Saxo Bank will have around 13–14 foreign sales offices, up from eight last year, Lars Seier Christensen told Dow Jones in May 2009[22] . In August 2008, Saxo Bank hired the former head of the facility service and cleaning company ISS A/S Eric Rylberg, as new Chief Operating Director, and Karsten Poulsen as his Deputy. The founders retained their executive titles. Shortly after, Rylberg and Poulsen implemented a rationalisation plan, reducing staff by around 1/3.[23]

Saxo has since been growing through add-on acquisitions and startups, and has built an asset management department, Saxo Asset Management, legally segregated from the rest of the Bank, through acquisitions. In 2009, Saxo acquired Sirius Kapitalforvaltning and in June 2009, it purchased the full share capital of Capital Four Management Fondsmaeglerselskab A/S and 51% of the share capital of Global Evolution Fondsmaeglerselskab A/S, providing expertise for Danish bond & equity instruments. According to its annual results for 2009, Saxo Asset Management grew assets under management from DKK 10 DKK billion to more than DKK 20 billion by the end of 2009 and another 25% in Q1 2010 to DKK 25 billion. A special report on Danish HNW clients in the New York Times said that by July 2010, Saxo Asset Management handles 29 billion kroner in assets, 15.5 billion of which comes from ultra-high-net-worth clients.[24]

Following a minority stake purchase in EuroInvestor, a Danish investment portal, in 2009,[25] Saxo said it would speed up development of its trading systems by acquirering a 40% stake in Initto [26], a software company based in India and Ukraine[27] and in 2010 it launched an equity profile incorporating elements from EuroInvestor. In July 2010, Saxo Bank increased its equity interest in EuroInvestor to 41%[28] .

On 4 December 2009, TV2 (Denmark)'s financial news disclosed that the founders of Saxo was announcing to E*TRADE's Swedish and Danish employees that Saxo had entered into an agreement to acquire E*TRADE International's local Nordic online trading business and online bank from E*TRADE Nordic AB, an indirect subsidiary of US-based E*TRADE FINANCIAL Corporation. In an interview with Dagens Industri, Lars Seier Christensen said that following regulatory approval Saxo Bank will have full banking license in Sweden. He also told its asset under management will increase by more than SEK 6.0bn and adds an additional 50,000 active accounts. E*TRADE's Nordic business also includes client accounts in the baltic region.[29] According to other reports Saxo will now be able to offer pension products, stock and margin accounts, bond offerings and Funds Supermarket.[30] [31] [32] Following a record year in 2008[3] , net profit ended at DKK 201 million in 2009 with 80% earned in the second half of 2009[4] .

On 6 October 2008, during the Financial Crisis, Saxo decided to join the Danish Sector Fund (Det Private Beredskab[33]), and was as such encompassed by the two-year guarantee for deposits with Danish financial institutions. Saxo Bank declined to participate in the Government's Credit Package[34] , saying it would lose money if it received capital from the state-funded Credit Package[35] . On 24 March 2010, a majority in the Danish Parliament agreed on a new guarantee scheme, which would bring deposit guarantees into line with European Union rules. The new scheme takes effect from 1 October 2010. Like all other Danish banks, Saxo Bank is covered by the scheme[36] .

2010 growth

In March, 2010, Saxo Bank launched a new equity platform able to trade in over 11,000 stocks and other products listed on 23 major exchanges worldwide.[37]

Saxo Bank was later named the *Best White Label Solution Provider* in the world, at the World Finance Foreign Exchange Awards 2010, citing "the Bank's long established program of investment and innovation within its White Label business".[38] [39] . In the same month the bank was also awarded *Best re-labelling platform for 2010* by industry magazine *Profit & Loss*. Saxo was voted into the top spot by subscribers to represent 'reader's choice'.[40]

In July, 2010, CEO Kim Fournais stated that the bank's 2009 turnover of 70 billion DKK on a daily basis had been broken, despite negative press at the start of the year.[41] The growth is believed partly to spring from an explosive growth in Saxo Bank's largest partner, Citi, whose turnover resulting from trading grew nearly 500% in the first 6 months of 2010. In August, 2010, the Bank's half-year results revealed that net profit had indeed increased 13-fold from 41 million DKK to 551 million DKK, making the first six months of 2010 better than any other full year in Saxo Bank's history.[5]

On 17 September 2010, Saxo Bank announced plans to invest 59 million DKK in Brorup Sparekasse; a small Danish savings bank founded in 1897 with seven branches.[42] The bank also said that it intended to offer at least 150 million DKK more in capital contribution when the terms for transforming the Brorup Sparekasse into a joint-stock company were fulfilled. Brorup Sparekasse ran into difficulties in 2008 and 2009, primarily due to failed speculation in unlisted bonds, which led to 2009 negative revenue of 74 million DKK before tax.[43] [44] Saxo Bank later announced that the terms for transforming Brorup Sparekasse into a joint-stock company were fulfilled with Saxo Bank as the largest shareholder owing 98,6%[45]

In November, 2010, Saxo Bank launched an online equity platform with Microsoft. The personal *MSN Trader – Powered by Saxo Bank* allows individual investors to trade 11,000 shares on 23 exchanges globally "in combination with functionalities which were previously available to professional investors only". These include in-depth research such as being able quickly to compare investment opportunities in a company with its peers in the same sector. The platform was tailor-made for the MSN audience, featured on MSN Money. It is said to mark one of the most significant deals in Saxo Bank's history.[8] [46]

In its 2010 results, Saxo Bank proved to have grown way beyond industry averages with profits trebling compared to 2009, but it also revealed that it had launched another online platform this time with Barclays Stockbrokers Ltd. to supply their UK client base with an International Equities solution. The solution, which goes under the name *International Trader*, encompasses over 9,000 equities, 21 major global exchanges in 13 international markets and an Equities research solution. The release were already being marketed widely in the UK marketplace as well as to Barclays' existing UK clients base.[47] In May, 2011, Saxo Bank furthermore announced it would provide TD Waterhouse, the UK's leading execution only broker, with an online derivatives trading platform for retail investors.[48]

Further expansion of products and services

In Scandinavia, Saxo Privatbank, a new physical branch in Hellerup, Copenhagen, Denmark, was launched offering banking services and advice to retail clients. Saxo Bank said that the newly opened branch will align its efforts with the specific wishes and needs of the clients and targets its offers towards clients who have their finances under control.[49] According to a spokeperson from the Bank, Saxo Privatbank will in due time be represented in a string of the bigger Danish cities.[50]

Saxo Bank tightened and standardised the pricing of foreign exchange options on its retail trading platform, SaxoTrader in a bid to boost transparency and attract more investors to the market. In an upgrade that went live in mid-May, the Bank improved spreads on FX options for all 40 of its currency pairs, with the new prices streamed to SaxoTrader. The tightening of spreads was applied to all existing and new clients, and was accompanied by an account standardisation process. Clients can now choose a classic, premium or platinum account, each with different spread levels and qualification criteria.[51]

Furthermore, Markit announced that Saxo Bank would become the first financial institution to report its CFDs on Single Stock trades on a voluntary basis, in a bid to bring greater transparency to this fast growing market.[52]

In order to attract more flow from high-frequency traders, Saxo Bank launched ForexTrading.com, a retail foreign exchange platform. Liquidity comes from the same 12 tier-one banks as for the Bank's SaxoTrader platform but the new service has a slimline offering, variable spreads, a low initial deposit of $2,000 and is serviced in English only.[53]

Corporate headquarters and architecture

Saxo Bank is headquartered in Tuborg Havn in the northern outskirts of Copenhagen. The building, which received an RIBA International Award, was designed by Danish architects 3XN, and is owned by Carlsberg Properties.[54] The façade of the building is covered in diagonal white aluminium and blue glass patterns.[55] In June 2010, Danish business newspaper, Borsen, named the Saxo headquarters the "Kingdom's finest domicile"[56]

Other Saxo Bank offices are located in Australia, the Czech Republic, Dubai, France, Greece, Hong Kong, Italy, Japan, Netherlands, Singapore, Spain, Switzerland and the United Kingdom.

Ownership and business model

Saxo Bank is privately held with co-founders Kim Fournais and Lars Seier Christensen holding majority stakes. Private Equity firm General Atlantic bought a minority stake in June 2005, for which the CEOs received more than $30 million each.[57] . According to the 2009 annual report, Fournais Holding A/S and Lars Seier Christensen Holding A/S today own 30% each of the Bank. General Atlantic owns 22% of the Bank and Espirito Santo Financial Group (Portugal's third biggest financial group) with its subsidiary Banco Espirito Santo Group together hold 9.9% of the Bank's shares, after taking a five percent stake in Saxo in 2008. The remaining interest is mainly held by a number of current and former employees of Saxo Bank.

According to Saxo representatives, so-called White Label Partnerships accounts for more than half of the bank's revenue providing re-distribution of the bank's platform to a broader public through partnerships. Private Wealth Management is a relatively new area of business.

Public relations

In 2008, Saxo Bank admitted editing its own Wikipedia entry through an anonymous user to remove information it said was untrue. When confronted by the media it acknowledged that a "slightly more elegant solution" should have been found.[58]

Sponsorships

In mid-June 2008, Riis Cycling A/S announced that Saxo Bank had entered a three year contract as title sponsor, with immediate effect, so the team entered the 2008 Tour de France as Team CSC Saxo Bank.[59] Carlos Sastre won the Tour, and the team took the team classification. In 2009 and 2010, Andy Schleck finished the Tour in second place. When the bank signed the sponsorship agreement with Riis Cycling A/S, it noted that the team "has the international reach and name recognition that means we will be able to get our message our to most of our client groups around the globe. We love the sport, and believe that together we will be winners.". While the bank announced in January 2010 that it would not extend its sponsorship beyond 2010[60] , it surprised everyone Lars Seier Christensen together with Bjarne Riis announced to the public on 3 August 2010 that Saxo Bank had extended its sponsorship of Riis Cycling for another year, and that Tour winner Alberto Contador had signed onto the team, with the aim of winning all Grand Tours (the Tour de France, the Giro d'Italia and the Vuelta a España) in one year[61] .

Saxo is a sponsor of 'CEPOS University', the company also pays for reprinting and distributing several books Denmark; including Ayn Rand's novel *Atlas Shrugged*, which the bank had retranslated into Danish and distributed 10,000 copies (each new employee of the bank receives a copy). In May 2007, Saxo Bank donated one million DKK to the extreme right party *Ny Alliance* (later the*Liberal Alliance*), CEO Lars Christensen later joined the party himself[62] .

2010 media crisis: allegations of misconduct and subsequent acquittal

Background

While the first half of 2010 had been months of tremendous growth for Saxo Bank, allegations from former employees and two former clients sparked a two month media crusade against the bank. Saxo Bank was later cleared of all allegations and Danish Berlingske Tidende characterised Saxo Bank's media crisis as a bitter clash between the founders and former employees as well as former clients most likely due to personal disappointments or economic matters[63] .

Media storm and subsequent steps by Saxo

In April 2010, a Portuguese financial institution accused Saxo Bank of manipulating stock and currency prices, claiming it lost 10 million euros.[64]

In May, a Lebanese investor said his company had lost 20 million DKK in four months because of unstable pricing and error interruptions on the SaxoTrader. Saxo Bank rejected that errors are a common thing and claimed that the SaxoTrader has an uptime of 100%[65] [66]

J. Korsoe Jensen, Attorney at Law for Saxo, issued lawsuits against the two former clients in June, 2010, for false accusations.[67] [68]

The new Chairman of the Board of Saxo Bank, Kurt Larsen, admitted to Berlingske Tidende that the lawsuits filed against the two former clients were indeed very unusual measures.[69] However, he also said:

> "We have tried, but have not been successful in gaining clients authorisation to respond to the accusations they have made against us. Therefore, we had to react, and therefore we have filed lawsuits against them."[69]

Completion of FSA review and independent investigation

In July 2010, the Danish FSA completed their investigation of Saxo Bank, having analysed its trading platform and protection of investors. The Danish FSA found its automated trading system sufficiently transparent to users. Furthermore, the FSA found no remarks vis-à-vis the bank's service regarding of trading tips on the platform, the technical structure of the system or its robustness. Likewise, the Danish FSA concluded that the evaluation version of Saxo Bank's trader would permit possible clients to experience dealing with trading complex products without risk before they became customers for real.

However, the Danish FSA ordered an independent investigation of whether trades executed manually and performed on the Bank's electronic trading system are generally being executed in accordance with the Bank's common trading conditions and Best Execution Policy.[70] In late August, it was announced by the Danish FSA that the investigation would be done by Oliver Wyman, the international consulting company.[71]

The independent investigation is said to have cost Saxo Bank millions and were two months overdue when on 29 November 2010, the Danish FSA published a statement, saying that Oliver Wyman had concluded that "Saxo Bank was not systematically mispricing its manual order flow to the detriment of its clients or in violation of its General Business Terms and its Best Execution Policy" and "that it could find no evidence of clients being systematically unfairly treated at the individual level". Based on the above mentioned conclusions, the Danish FSA found that the investigation by Oliver Wyman did not give cause for pursuing anything further regarding Saxo Bank's execution of manually executed trading orders nor regarding Saxo Bank's trading system. The statement also said that the instruments represented in the cases covered by the Danish media, this summer, and the client complaints received by the Danish FSA had been taken into account.[72]

Criticism of the Danish FSA

Following the announcement of the Danish FSA on 8 July, Lars Bo Langsted, a professor of economic law at Aalborg University stated that:[73]

> "The Danish FSA has become more aggressive. It has become quite common that the FSA reports companies to the police even though their violations aren't that serious. Where previously they might have chosen a dialogue, today their finger is on the trigger."[69]

The WeeklyFix [74] column suggested that the Danish FSA had been pressured to act on Saxo by press reports. "There is no doubt the Danish FSA has felt under pressure to act (...) This begs the question, since when did financial authorities spring into action on the back of media speculation?"[75]

References

[1] http://www.saxobank.com/

[2] http://www.investorwords.com/588/brokerage_house.html

[3] "Saxo Bank announces record profits" (http://www.eyeofdubai.com/v1/news/newsdetail-29258.htm) (in (Danish)). Eyeofdubai.com. 9 March 2010. . Retrieved 3 September 2010.

[4] "Page Unavailable – MSN Money" (http://news.moneycentral.msn.com/provider/providerarticle.aspx?feed=PR&date=20100406&id=11375116) (in (Danish)). News.moneycentral.msn.com. . Retrieved 3 September 2010.

[5] "Saxo's profit up 13-fold, 30% may be for sale" (http://forexmagnates.com/saxos-profit-up-13-fold-30-may-be-for-sale/?utm_source=rss&utm_medium=rss&utm_campaign=saxos-profit-up-13-fold-30-may-be-for-sale). forexmagnates.com. . Retrieved 19 August 2010.

[6] Af David Bentow. "Storaktionær i Saxo Bank overvejer salg af aktiepost – Finans" (http://www.business.dk/finans/storaktionaer-i-saxo-bank-overvejer-salg-af-aktiepost) (in (Danish)). Business.dk. . Retrieved 3 September 2010.

[7] Af Kasper Kronenberg. "Kapitalfond sælger Saxo-aktier inden sommerferien" (http://finanswatch.dk/Finansnyt/Pengeinstitutter/article2358801.ece) (in (Danish)). Finanswatch.dk. . Retrieved 4 March 2011.

[8] http://www.ft.com/cms/s/0/631cd63e-ea9a-11df-b28d-00144feab49a.html#axzz14n7TN8A0

[9] http://forexmagnates.com/saxo-trebles-profits-in-2010-signs-barclays-as-a-white-label/

[10] http://www.foxbusiness.com/industries/2011/05/18/saxo-bank-provide-derivatives-platform-td-waterhouse

[11] Milliardærer fra sidegaden, Berlingske Tidende, 21 June 2005

[12] *Economic Weekly*, 12–97

[13] *Economic Weekly*, 19–97

[14] *Economic Weekly*, 25–97, ff

[15] *Economic Weekly*, 7 April 1997

[16] "Saxo Bank Outsourcing, Online Trading Platform, MiniTrader" (http://web.archive.org/web/20080309121213/http://www.reksoft.com/customers/case-studies/saxo/). Reksoft.com. Archived from the original (http://www.reksoft.com/customers/case-studies/saxo/) on 9 March 2008. . Retrieved 3 September 2010.

[17] "SAXO BANK: Providing maximum flexibility through multi-asset trading platforms" (http://www.theneweurope.co.uk/feature-237/saxo-bank-providing-maximum-flexibility-through-multi-asset-trading-platforms.html). The New Europe. 1 December 2010. .

[18] Olivari, Nick (20 November 2007). "20 Nov (Reuters) – Citigroup Inc. (C.N), the largest U.S. bank, and Saxo Bank, the Copenhagen-based online investment bank, on Tuesday announced plans to launch CitiFX Pro, an online foreign exchange trading platform" (http://www.reuters.com/article/idUSN2056201120071120). Reuters. . Retrieved 3 September 2010.

[19] Af Birgitte Erhardtsen, Opdateret 22.00 af Kim Bové. "Saxo Bank afviser at Citibank-samarbejde skuffer – Finans" (http://www.business.dk/finans/saxo-bank-afviser-citibank-samarbejde-skuffer) (in (Danish)). Business.dk. . Retrieved 3 September 2010.

[20] FXweek, 27 September 2004

[21] FXweek, 20 February 2006, Vol 17 No 8

[22] INTERVIEW: Saxo Eyes Asset Mgmt Add-On Buys, Opens Offices By Anna Molin Of DOW JONES NEWSWIRES, 13 May 2009

[23] "Saxo Bank fyrer hver tredje i Danmark – Finans" (http://www.business.dk/article/20080922/finans/80922067/) (in (Danish)). Business.dk. . Retrieved 3 September 2010.

[24] Rachlin, Natalia (16 July 2010). "Cradle-To-Grave System Gets Rocky in Denmark" (http://www.nytimes.com/2010/07/16/business/global/16iht-nwdenmark.html?pagewanted=1). *The New York Times*. .

[25] "Saxo buys into EuroInvestor.com February weeklyFiX 2009. FX news, FX comment, FX hires, FX poll" (http://www.euromoneyfix.com/Article.aspx?ArticleID=2095990&PageID=8248). Euromoneyfix.com. 5 February 2009. . Retrieved 3 September 2010.

[26] http://www.initto.com/

[27] "Saxo Bank acquires a 40 per cent stake in Initto" (http://www.bankingtech.com/bankingtech/article.do?articleid=20000163222). *Banking Technology magazine*. . Retrieved 3 September 2010.

[28] "Saxo Bank øger ejerandel i Euroinvestor" (http://epn.dk/brancher/finans/article2114474.ece) (in (Danish)). Epn.dk. 15 July 2010. . Retrieved 3 September 2010.

[29] "Saxo Bank köper nordiska E-trade" (http://di.se/Avdelningar/Artikel.aspx?ArticleID=2009\12\03\364702&sectionid=Ettan) (in (Danish)). DiSe. 3 December 2009. . Retrieved 3 September 2010.

[30] . http://online.wsj.com/article/BT-CO-20091203-702929.html.

[31] Af David Bentow. "Saxo Bank-køb af Etrade i Norden skal styrke salg af pension og aktier – Finans" (http://www.business.dk/finans/saxo-bank-koeb-af-etrade-i-norden-skal-styrke-salg-af-pension-og-aktier) (in (Danish)). Business.dk. . Retrieved 3 September 2010.

[32] "Saxo Bank-boss: Har flere ting i ærmet" (http://borsen.dk/investor/nyhed/171614/) (in (Danish)). Borsen.dk. . Retrieved 3 September 2010.

[33] "Det Private Beredskab – Finansrådet" (http://www.finansraadet.dk/danish/menu/omkundeforhold/detprivateberedskab). Finansraadet.dk. . Retrieved 24 September 2010.

[34] "Credit package agreement in place – Økonomi- og Erhvervsministeriet" (http://www.oem.dk/sw24128.asp). Oem.dk. 30 June 2009. . Retrieved 24 September 2010.

[35] "Ja til bankpakke II ville give tab til Saxo Bank" (http://borsen.dk/finans/nyhed/160922/newsfeeds_rss) (in (Danish)). Borsen.dk. . Retrieved 3 September 2010.

[36] "Denmark halts bank account guarantee" (http://www.swedishwire.com/nordic/3492-denmark-halts-bank-account-guarantee) (in (Danish)). Swedishwire.com. . Retrieved 3 September 2010.

[37] "Saxo Bank launches new equity platform" (http://www.bankingtech.com/bankingtech/article.do;jsessionid=8004375DBDD5EEDC401AD3203EB41ED8.5d25bd3d240cca6cbbee6afc8c3b5655190f397f?articleid=20000175484). *Banking Technology*. . Retrieved 3 September 2010.

[38] "Saxo Bank named "Best White Label Solution Provider" in World Finance's FX Awards" (http://www1.albawaba.com/en/news/saxo-bank-named-best-white-label-solution-provider-world-finances-fx-awards) (in (Danish)). Al Bawaba. 30 June 2010. . Retrieved 3 September 2010.

[39] "SaxoTrader named Best Re-Labelling Platform by business magazine Profit & Loss" (http://www.zawya.com/story.cfm/sidZAWYA20100713110030/SaxoTrader named Best Re-Labelling Platform by business magazine Profit &Loss). Zawya.com. . Retrieved 3 September 2010.

[40] "Forex brokers news – SaxoTrader named Best Re-Labelling Platform by business magazine Profit & Loss" (http://www.forex-ratings.com/brokers-news/?id=4014). Forex-ratings.com. . Retrieved 3 September 2010.

[41] "Saxo Bank tjener som aldrig før" (http://epn.dk/brancher/finans/bank/article2105037.ece) (in (Danish)). Epn.dk. 15 July 2010. . Retrieved 3 September 2010.

[42] "Brorup Temps Saxo With Ready Systems" (http://borsen.dk/nyheder/finans/artikel/1/191448/broerup_frister_saxo_bank_med_koereklare_systemer.html) (in (Danish)). Borsen.dk. 17 September 2010. . Retrieved 17 September 2010.

[43] "Saxo set to take over Jutland bank" (http://www.cphpost.dk/business/119-business/50018-saxo-bank-set-to-take-over-jutland-bank.html). http://www.cphpost.dk. . Retrieved 17 September 2010.

[44] "Saxo Bank Wants To Be Largest Shareholder in Brorup Sparekasse with a Capital Contribution Exceeding 200 Million DKK" (http://www.prnewswire.com/news-releases/saxo-bank-wants-to-be-largest-shareholder-in-brorup-sparekasse-with-a-capital-contribution-exceeding-200-million-dkk-103149029.html). http://www.prnewswire.com. . Retrieved 17 September 2010.

[45] "Saxo Bank want to enter the big cities" (http://www.finansforbundet.dk/default.asp?mId=2071&ArtId=178885). Jyllands-Posten. . Retrieved 11 November 2010.

[46] "New MSN-branded trading platform from Saxo Bank is aimed at retail investors" (http://bankingtech.com/bankingtech/new-msn-branded-trading-platform-from-saxo-bank-is-aimed-at-retail-investors/20000191741.htm;jsessionid=04C626418DB9763FC28EF3C02F92DAFD.f11b1cefac76ad95c7627468fee9bde7e866d022). http://www.bankingtech.com. . Retrieved 11 November 2010.

[47] "Saxo trebles profits in 2010, signs Barclays as a White Label!" (http://forexmagnates.com/saxo-trebles-profits-in-2010-signs-barclays-as-a-white-label/). http://forexmagnates.com. . Retrieved 3 March 2011.

[48] "Saxo Bank Strikes Again" (http://www.forexstreet.net/profiles/blogs/saxo-bank-strikes-again-white). forexstreet.net. . Retrieved 18 May 2011.

[49] "Saxo Privatbank Opens First Branch for Retail Market" (http://ca.finance.yahoo.com/news/Saxo-Privatbank-Opens-First-prnews-3332252963.html?x=0). Yahoo Finance Canada. . Retrieved 9 May 2011.

[50] "Saxo Privatbank opens new physical branch for retail clients in Copenhagen". Retail Banking Business Review.

[51] "Saxo steps up options pricing on SaxoTrader" (http://www.fxweek.com/fx-week/news/2073560/saxo-steps-options-pricing-saxotrader). FX Week. . Retrieved 30 May 2011.

[52] "Markit to Provide First CFD Trade Reporting System to Saxo Bank" (http://www.reuters.com/article/2011/06/09/idUS71885+09-Jun-2011+PRN20110609). Reuters. 9 June 2011. . Retrieved Thu 9 Jun 2011.

[53] "Danish bank says it wants to cater for 'more self-driven' clients" (http://www.fxweek.com/fx-week/news/2075663/saxo-targets-traders-forextradingcom). FX Week. . Retrieved 6 June 2011.

[54] "Winners 2009" (http://www.architecture.com/Awards/RIBAInternationalAwards/2009/Winners2009.aspx). RIBA. . Retrieved 15 September 2009.

[55] "Saxo Bank wins RIBA International Award" (http://www.worldarchitecturenews.com/index.php?fuseaction=wanappln.projectview&upload_id=11783). World Architecture News. . Retrieved 15 September 2009.

[56] "Saxo Bank: Rigets flotteste domicil?" (http://pleasure.dk/design/artikel/183473/saxo_bank_rigets_flotteste_domicil.html) (in (Danish)). Pleasure.dk. 3 June 2010. . Retrieved 3 September 2010.
[57] Saxo founders cash in: FXweek, Vol 16 No 25
[58] "Saxo Bank afviser propaganda på internettet" (http://borsen.dk/nyheder/it/artikel/1/141222/saxo_bank_afviser_propaganda_paa_internettet.html) (in (Danish)). Borsen.dk. . Retrieved 3 September 2010.
[59] "Riis Cycling signs Sponsorship Agreement with Saxo Bank" (http://www.riis-cycling.com/ny_news.asp?n_id=1918&lang=uk). Riis-cycling.com. 1 January 2009. . Retrieved 3 September 2010.
[60] "Saxo Bank to end team sponsorship" (http://www.cyclingnews.com/news/saxo-bank-to-end-team-sponsorship). Cyclingnews.com. 8 January 2010. . Retrieved 3 September 2010.
[61] "OFFICIELT: Saxo Bank fortsætter – Contador ny mand | TV 2 SPORT" (http://tv2sport.dk/node/60255) (in (Danish)). Tv2sport.dk. . Retrieved 3 September 2010.
[62] "Bankdirektør melder sig ind i NA" (http://politiken.dk/politik/article541715.ece) (in (Danish)). Politiken.dk. 27 August 2010. . Retrieved 3 September 2010.
[63] Af Jens Chr. Hansen. "» Handler balladen i Saxo Bank om et medarbejder- og ejeropgør? – Børsen & bosserne" (http://bosserne.blogs.business.dk/2010/04/20/handler-balladen-i-saxo-bank-om-et-medarbejer-og-ejeropgÃ¸r/) (in (Danish)). Bosserne.blogs.business.dk. . Retrieved 3 September 2010.
[64] "Saxo Bank founders accused of fraud" (http://www.cphpost.dk/business/119-business/48865-saxo-bank-founders-accused-of-fraud.html). Copenhagen Post. 29. . Retrieved 24 September 2010.
[65] Af Peter Suppli Benson, Birgitte Erhardtsen. "Saxo-kunde taber 20 mio. på fire måneder – Finans" (http://www.business.dk/finans/saxo-kunde-taber-20-mio.-paa-fire-maaneder) (in (Danish)). Business.dk. . Retrieved 3 September 2010.
[66] "Saxo Bank afviser anklager om mange fejl" (http://borsen.dk/finans/nyhed/184224/) (in (Danish)). Borsen.dk. . Retrieved 3 September 2010.
[67] "Saxo Bank suing Fortune and IM International" (http://www.fxstreet.com/news/forex-news/article.aspx?storyid=bda15824-5908-4584-a47b-cf320e5fb783). FXStreet. 7. . Retrieved 24 September 2010.
[68] Af Sonny Wichmann. "Saxo Bank anlægger sag mod kritiske kunder – Finans" (http://www.business.dk/finans/saxo-bank-anlaegger-sag-mod-kritiske-kunder) (in (Danish)). Business.dk. . Retrieved 3 September 2010.
[69] Af Peter Suppli Benson. "Error: no |title= specified when using {{Cite web}}" (in (Danish)). Berlingske Tidende.
[70] Greenberg, Michael. "Independent investigation costs Saxo Bank millions" (http://forexmagnates.com/independent-investigation-costs-saxo-bank-millions/). Forex Magnates. . Retrieved 24 September 2010.
[71] Af David Bentow, RB-Børsen. "International konsulent skal undersøge Saxo Bank – Finans" (http://www.business.dk/finans/international-konsulent-skal-undersoege-saxo-bank) (in (Danish)). Business.dk. . Retrieved 3 September 2010.
[72] Greenberg, Michael. "Saxo cleared of market prices manipulation" (http://forexmagnates.com/saxo-cleared-of-market-prices-manipulation/). Forex Magnates. . Retrieved 30 November 2010.
[73] *Politiken*, 10 July 2010, section 1, page 11
[74] http://www.euromoneyfix.com/
[75] *The WeeklyFix*, 23 July 2010

External links

- www.saxobank.com (http://www.saxobank.com)

Matchbook_FX

Matchbook FX	
Type	Online Foreign Exchange ECN
Location	New York, NY, US

Matchbook FX was an internet-based Electronic communication network ("ECN" or Electronic Trading Network) for trading currency online in the FX or Foreign exchange market.

History

Founded in 1999, Matchbook FX (sometimes referenced as MatchbookFX) was the world's first [1] [2] open and inclusive internet ECN for Foreign Exchange trading[3] , available to all willing FX trading participants including hedge funds, CTAs, banks, corporations and, uniquely at the time, retail FX traders as well[4] .

Matchbook FX was initially conceived by Daniel Uslander, Ron Comerchero (commodity futures and equities traders) and Josh Levy (former Goldman Sachs FX trader) of the New York based proprietary FX trading firm Valhalla Forex Inc. and officially formed in April 1999 as a partnership between Valhalla Forex and the Florida based equities-trading technology firm NexTrade ECN [5]. Several months later, GlobalNetFinancial.com, a NASDAQ-traded financial news and technology firm, bought in as the third major equity partner in the three-way joint venture.[6]

As one of the earliest providers of open-access FX e-trading, Matchbook FX received considerable acclaim for its efforts to instigate change and level the playing field in the insular, closed, clubby and highly profitable domain of interbank Forex dealing, likely to the chagrin of the major international money center banks[7] . Matchbook FX was recognized in 2000 as one of *Silicon Alley Reporter* Magazine's "12 to Watch", its annual listing featuring top internet companies.[8]

Unique aspects

Prior to Matchbook FX, most FX trading was transacted mainly by phone or amongst large banks (such as Chase, Goldman Sachs, UBS, Deutsche Bank or Citibank) in the "interbank market" or by phone between large banks and their multinational corporate clients (such as IBM, Intel, Coca Cola, etc.) or institutional clients (such as hedge funds, pensions, mutual funds and other asset managers).

Despite on-line FX "e-trading" being rare at the time, Matchbook FX's approach was also considered unique by market participants because of its stated aims to democratize the Foreign Exchange market[9] by empowering all "Buy side" FX participants (including retail traders and institutions) to be, for the first time, "Price-Makers" instead of only "Price-Takers."[10] Matchbook FX functioned as an open limit order-book, similar to an online exchange, where any participant subscribed to the network could either post its own bids and offers just like a market maker, or immediately trade on any other existing bids and offers for a given currency. This process allowed users to join or better the prevailing prices in the network and thus directly impact (and tighten) the bid/ask spread widths on which they traded. As such, Matchbook FX was considered to be one of the main catalysts that presaged rapid technological advances, sharp compressions in bid/ask spreads and other sweeping changes into the currency market.

Matchbook FX was also unique in that, in addition to its own user-generated bids and offers, it was the first ever e-FX trading platform to feature dynamic fully automated, executable streaming currency prices (as opposed to Request for Quotation "RFQ" driven prices) contributed directly from a major FX bank (rumored to have been Deutsche Bank). Though such a configuration is now commonplace, it was widely considered by those in the industry to be a technological innovation in 1999.

Matchbook FX also innovated on the US FX regulatory front, by being the first major US-based FX dealer to voluntarily subject itself to NFA (National Futures Association) regulations,[11] prior even to the Commodity Futures Modernization Act of 2000. At the time of Matchbook FX's launch, a handful of other retail-focused online FX dealers existed, none of which were NFA regulated, notably CMC Markets, GFT Forex and MG Forex [12] but these firms served as market-makers where clients could only trade on the prices they were streamed, at bid/ask spreads they could not control. Following Matchbook FX, many technology-driven FX trading operations soon emerged including Currenex, Atriax (now defunct), FXall [13], Hotspot FX (a Matchbook FX carbon-copy ECN), as well as FXCM, Gain Capital and Saxo Bank which together with CMC Markets ushered in an explosive new era of online retail FX trading.

Final days

The implosion of the dot-com bubble in 2000 severely hindered Matchbook FX's ability to raise continued operating funds. The stock price of GlobalNetFinancial.com (NASDAQ: GLBN), Matchbook FX's main financial backer, fell from a high of over $70 per share in 2000 to less than $1 per share by the end of 2001. By 2002, Matchbook FX was sold to a consortium of investors who made the decision to discontinue the Matchbook FX brand.

References

[1] Waters Magazine - Special Report on Best Execution May 1st, 2001 http://www.watersonline.com/public/showPage.html?page=271702

[2] Waters Online SPECIAL REPORT: FOREIGN EXCHANGE IT - 1 May 2001 http://www.watersonline.com/public/showPage.html?page=271704

[3] Euromoney Magazine Cover Article A True Exchange for Forex June 2000 http://www.euromoney.com/Article/1008094/E-forex-A-true-exchange-for-forex.html

[4] *The Washington Post*, August 12, 1999, Section A; p. A01. Dugan, Ianthe Jeanne, "Do-It-Yourself Traders' Riskiest Purchase: Currency," http://pqasb.pqarchiver.com/washingtonpost/access/43884076.html?FMT=ABS&FMTS=ABS:FT&date=Aug+12%2C+1999&author=Ianthe+Jeanne+Dugan&desc=Do-It-Yourself+Traders%27+Riskiest+Purchase%3A+Currency

[5] http://www.nextrade.com

[6] GlobalNet, NexTrade ECN and Valhalla Forex Announce Launch of Matchbook FX, LLC. http://www.prnewswire.co.uk/cgi/news/release?id=51825

[7] *FX Week Magazine*, "New Web Trading Service Tightens Squeeze on Traditional FX Business", October 18, 1999 | Vol 10 No 41 http://www.tactical-group.com/1999.10.18-FXweek.htm

[8] MatchbookFX Named One of Silicon Alley Reporter's Silicon Alley Reporter "12 to Watch"; Annual Listing Features Top Internet Companies of 2000. http://www.highbeam.com/search.aspx?q=MatchbookFX%20Named%20One%20of%20Silicon

[9] *The Economist Magazine*, August 19, 2000, http://www.oanda.com/corp/news/media/economist_08212000

[10] The New Future of Forex Trading http://onlinetradingstockandoption.net/the-future-of-forex-trading

[11] NFA Matchbook FX Registration Record http://www.nfa.futures.org/BasicNet/Details.aspx?entityid=0297892&rn=N

[12] http://www.mgforex.com

[13] http://www.FXall.com

External links

- *The Economist Magazine*, August 19, 2000 (http://www.oanda.com/corp/news/media/economist_08212000)
- *Euromoney Magazine*, Cover article, June 2000 (http://www.tactical-group.com/2000.6_EuromoneyArticle.htm)
- *Economist Magazine*, August 19, 2000 (http://www.highbeam.com/doc/1G1-64361579.html)
- (http://www.nextrade.com) NexTrade Holdings, Inc. / NexTrade ECN http://www.nextrade.com
- (http://www.smartmoney.com/investing/stocks/A-New-Refuge-for-Day-Traders-7584/) *SmartMoney Magazine* April 27, 2000
- (http://www.highbeam.com/search.aspx?q=MatchbookFX Files for U.S.Patent) Matchbook FX Files for U.S. Patent for Currency Trading Platform

MetaTrader_4

Trading Platform
MetaTrader 4

Developer(s)	Metaquotes Software Corp.
Operating system	Windows
Type	Trading platform, Technical analysis software
License	Metaquotes Software Corp.
Website	www.metaquotes.net/en/metatrader4 [1]

MetaTrader 4, also known as **MT4**, is an electronic trading platform widely used by online retail foreign exchange speculative traders. It was developed by MetaQuotes Software and released in 2005. The software is licensed to foreign exchange brokers who provide the software to their clients. The software consists of both a client and server component. The server component is run by the broker and the client software is provided to the brokers customers, which use it to see live streaming prices, charts and to place orders as well as manage their account.

The client is a Microsoft Windows-based application that became popular mainly due to the ability for end users to write their own trading scripts and robots that could automate trading. In 2010 MetaQuotes released a successor, **MetaTrader 5**, however uptake was slow and as of early 2012 most brokers still used MT4.

History

Its developer MetaQuotes Software had previously released a number of versions of the MetaTrader software starting in 2002. Metatrader 4 was a significantly enhanced version and was released in 2005.

Between 2007 and 2010 a number of brokerages added the MT4 platform as an optional alternative to their existing trading software due to its popularity with traders and the large number of third party scripts and advisors.

In 2010 a significantly re-coded MetaTrader 5 went into public beta testing and was subsequently released later in the year

Functionality

A chart from the MT4 trading screen

The client terminal includes a built-in editor and compiler with access to a user contributed free library of software, articles and help. The software utilizes a proprietary scripting language, MQL4[2] , which enables traders to develop **Expert Advisors**, custom indicators and scripts. MetaTrader's popularity largely stems from its support of algorithmic trading.

Yahoo hosts a large group (over 12,000 members) devoted to development of free open source software for MetaTrader.[3]

MT4 is designed to be used as a stand alone system with the broker manually managing their position and this is a common configuration used by brokers. However a number of third party developers have written software bridges enabling integration with other financial trading systems for automatic hedging of positions.[4]

Components

The complete MetaTrader 4 package includes the following components:

- **MetaTrader 4 Client Terminal** - the client part. Provided free by brokerages for real-time online trading and as Demo (practice trading) accounts. This provides trade operations, charts and technical analysis in real time. The internal C-like programming language allows users to program trading strategies, indicators and signals. 50 basic indicators are included, each of which can be further customized. The software runs on Windows 98/2000/XP/Vista/7. Some users have reported success using WINE on Linux for the client terminal.[5]
- **MetaTrader 4 Mobile** - controls a trading account via mobile devices such as mobile phones or PDAs. Runs on Windows Pocket PC 2002/Mobile 2003. Newly released mobile versions also run on the iPhone, iPad, and Android devices.
- **MetaTrader 4 Server** - the core of the system, the server part. Designed to handle user requests to perform trade operations, display and execution of warrants. Also, sends price quotes and news broadcasts, records and maintains archives. Works as a service. Does not have a separate interface.
- **MetaTrader 4 Administrator** - is designed to remotely manage the server settings.
- **MetaTrader 4 Manager** - designed to handle trade inquiries and manage customer accounts.
- **MetaTrader 4 Data Center** - a specialized proxy server and can be an intermediary between the server and client terminals. It reduces the price quote sending load on the main server.

Products

The platform is focused on margin trading. Some brokerages use the platform to trade CFD but it is not designed for full-time work in the stock market or to conduct non margined trading.

MetaTrader 5 is designed as a platform for equity markets as well as foreign exchange, CFD and futures trading.

Both MetaTrader 4 and 5 can use custom indicators and trading programs (called Expert Advisors) for trading automation.

See also

- Straight-through processing
- Electronic communication network

References

[1] http://www.metaquotes.net/en/metatrader4
[2] "MQL4 Documentation" (http://docs.mql4.com/). Metaqoutes Software. . Retrieved June 15, 2011.
[3] "Yahoo Groups - MetaTrader_Experts_and_Indicators" (http://finance.groups.yahoo.com/group/MetaTrader_Experts_and_Indicators/). Yahoo. . Retrieved June 15, 2011.
[4] "Google search - MT4 Bridges" (http://www.google.com.au/search?hl=en&q=MT4+bridges&btnG=Search&meta=&aq=f&aqi=&aql=&oq=&gs_rfai=). Google. . Retrieved June 17 2011.
[5] Jafar Calley (17 June, 2010). "Installing Metatrader 4 Forex Charts Under Linux" (http://www.aboutcurrency.com/university/metatrader/installing_metatrader4_under_linux.shtml). AboutCurrency.com. .

External links

- Trading Platform MetaTrader 4 (http://www.metaquotes.net/en/metatrader4)
- MetaQuotes Software Corp. Official website (http://www.metaquotes.net/)
- MetaTrader 5 Official website (http://www.metatrader5.com/)

MetaQuotes_Software

MetaQuotes SOFTWARE	
Type	Private
Industry	Software
Founded	2000
Headquarters	Limassol, Cyprus
Key people	Renat Fatkhullin CEO
Products	Trading systems; MetaTrader 4 MetaTrader 5 MetaTrader 4 Mobile MetaTrader 5 Mobile TeamWox
Employees	50+
Website	www.metaquotes.net [1]

MetaQuotes Software Corp is a software company that produces online trading platform *MetaTrader 4* more commonly known as **MT4**. It specializing in the development of software for financial markets and is registered in Cyprus but founded and primarily based in Russia. The companies main focus is on Forex market trading software with its trading platform MetaTrader being used in over 450 brokerage companies and banks all over the world.

History

MetaQuotes Software Corporation was established in 2000 in Russia with the launch of **FX Charts** an online charting package used for margin FX trading. It became popular with traders as it could be run on a standard PC running Windows and did not require dedicated and expensive equipment.

In 2001 it released its second major revision of application **MetaQuotes** which expanded the range of instruments that could be traded. In addition to currencies it also included support for the newly emerging CFDs (Contracts for difference). In addition, the application allowed users to write their own trading programs and strategies using its own MetaQuotes Language (MQL). This substantially increase the flexibility and utility of the client trading palatform.

In early 2002, it released its third revision **MetaTrader**. This again expanded the number of financial products with the addition of futures contracts, but more importantly it expanded programming language for trading strategies with MQL II. A number of users and third parties started to develop functions and libraries using the new version of the built in programming language and APIs.

A chart from the MT4 trading screen

On July 1, 2005, it released **MetaTrader 4** or **MT4**. This was a major overhaul of the platform. This version allowed the client trading platform to connect via an optional 'Data Center' which dramatically increase the server side stability, reduced server load and made it more resilient to security threats such as Denial-of-service attacks. The client software was also upgraded with a new version of the C-like language for trade strategies programming called MetaQuotes Language 4 (MQL 4). Later that year MetaQuotes also released its first version of the trading platform for mobile devices.

In 2005 development of the **TeamWox**, a CRM system started for the company internal use. In 2009 the first public commercial version of TeamWox was released and went on to win some industry awards.[2] [3] [4]

The MT4 communication protocol was not backwardly compatibility with previous versions and MetaQuotes Software refused to open the Client API meaning all of the previous client developed code no longer worked. The client API did however significantly extend the functionality of the programming language for trading strategies with MQL4 including support for a DLL interface. This version of the software became popular with a large number of users and became to main product for MetaQuotes Software for the next 5 years.

On June 1, 2010 MetaQuotes Software officially released **MetaTrader 5** or **MT5** the next version of its trading platform.[5] It expanded the number of products that could be traded from just forex to include stocks and commodities.[6] MT5 also includes a new version of the programming language called MQL5. The new MT5 client trading terminal differs in a number of ways from MT4 and was not backwardly compatible.[7] [8]

Uptake of the new version was slow and by April, 2011 MetaTrader only 50 companies had new version with over 400 remaining on MT4 mainly due to the lack of backward compatibility.[9]

On April, 2011 MetaQuotes launched a service it called **MQL5 Cloud Network** which allowed users to sell their idle time for distributed computing tasks.[10]

In 2011 MetaQuotes Software Corp. entered into a partnership with International Business Times to deliver integrated trading platform and newswire service.[11]

In 2012 MetaTrader 5 Trading Platform has Certified for the Singapore Mercantile Exchange.[12] Metaquotes Software has been granted ISV status by the stock exchange.[13]

Automated Trading Championship

Starting in 2006, the company started holding annual auto trading championships for software developers in MQL4 with the aim of demonstrating the possibilities of automated trading. This event see specially written robotic computer programs face off against each other by analyzing the prices of various financial instruments and trading these in real time. Under the terms of the championship all of the programs would run simultaneously for three months under identical conditions. The Automated Trading Championship has the official web site.[14]

In 2011 has launched the fifth annual Forex Trading Robots (Expert Advisors in MQL5) competition.[15] In different years the competition is sponsored by MIG Bank, FXCM, Dow Jones & Company and other companies.

Operations

The company is registered in the offshore zone in Cyprus from where it does sales, marketing and technical support. Software development is done from Russia. The Cyprus-based company also has representative offices in Turkey, China, Singapore, Australia and Brazil.

References

[1] http://www.metaquotes.net

[2] "Best Soft 2010" (http://pcmag.ru/reviews/sub_detail.php?ID=43162) (in ru). PC Magazine/RE. 2010-11-21. . Retrieved 2012-02-06.

[3] "Best Soft 2011" (http://www.pcmag.ru/reviews/sub_detail.php?ID=45662&SUB_PAGE=0#teamwox) (in ru). PC Magazine/RE. 2011-11-21. . Retrieved 2012-02-06.

[4] "2012 Best CRM Software" (http://crm-software-review.toptenreviews.com). . Retrieved 2012-02-06.

[5] "Retail e-FX Provider: MT5 - Evolution or Revolution?" (http://www.e-forex.net/articles/Retail+e-FX+Provider/1556/MT5+-+Evolution+or+Revolution3F+). e-Forex Magazine. 2010-10-01. . Retrieved 2012-02-07.

[6] "Exclusive interview with Alexander Saidullin, Director of Trading Platform Development at MetaQuotes" (http://www.fxtradermagazine.com/archive_jul10_en.php). FX Trader Magazine. 2010-07-01. . Retrieved 2012-02-06.
[7] "Retail e-FX Client: MT5 widening the horizon for all Forex traders" (http://www.e-forex.net/articles/Retail+e-FX+Client/1612/MT5+E28093+widening+the+horizon+for+all+Forex+traders). e-Forex Magazine. 2010-04-01. . Retrieved 2012-02-07.
[8] "New trading platform MetaTrader 5" (http://blog.instaforex.com/?p=562). InstaForex. 2010-10-21. . Retrieved 2012-02-06.
[9] "Interview with Metaquotes – Lenar Fatkhullin" (http://forexmagnates.com/interview-with-metaquotes-lenar-fatkhullin). Forex Magnates. 2011-04-29. . Retrieved 2012-02-07.
[10] "Statistics of the Distributed Cloud Network" (http://cloud.mql5.com/en). . Retrieved 2012-02-08.
[11] "IBTimes and MetaQuotes Partner to Deliver Integrated Trading Platform and Newswire Service" (http://forexmagnates.com/ibtimes-and-metaquotes-partner-to-deliver-integrated-trading-platform-and-newswire-service/). Forex Magnates. 2011-02-01. . Retrieved 2012-02-07.
[12] "SMX certifies MetaQuotes trading platform" (http://www.automatedtrader.net/news/exchange-news/104446/smx-certifies-metaquotes-trading-platform). Automated Trader. 2012-01-19. . Retrieved 2012-02-07.
[13] "Vendors with SMX" (http://www.smx.com.sg/Technology/VendorsWithSMX/Index.aspx). Singapore Mercantile Exchange. . Retrieved 2012-02-07.
[14] "Official website of the fifth annual Forex Trading Robots competition" (http://championship.mql5.com/). . Retrieved 2012-02-07.
[15] "Robots Trade Stocks as MetaQuotes' Competition Begins" (http://www.ibtimes.com/articles/223617/20111003/robots-forex-stocks-competition-metaquotes.htm). IB Times Staff Reporter. 2011-10-03. . Retrieved 2012-02-07.

External links

- Official website of MetaQuotes Software Corp. (http://www.metaquotes.net/en)
- Trading Platform MetaTrader 4 (http://www.metatrader4.com)
- Trading Platform MetaTrader 5 (http://www.metatrader5.com)
- MetaQuotes Language 4 (http://www.mql4.com)
- MetaQuotes Language 5 (http://www.mql5.com)
- MQL5 Cloud Network (http://cloud.mql5.com)
- Official website of the TeamWox system (http://www.teamwox.com/en)

Article Sources and Contributors

Retail_foreign_exchange_platform *Source*: http://en.wikipedia.org/w/index.php?title=Retail_foreign_exchange_platform *Contributors*: Aerozot, Amiodarone, Andharyel, Andrewelle1, Careerdaytrader, Dawnseeker2000, Dina Jones, Dxantos, Ether13, Excirial, FX2011, ForexAficionado, Fæ, Gurt Posh, Jacob Lundberg, Joelrussell, Kallios, Korath, Lmatt, Macaddct1984, MrOllie, Ohnoitsjamie, Patchy1, Raiden575, Robofish, Sargdub, Shoval55, Smallbones, Sridharan2010, Tabletop, Titus999, Tymewriter, Vicsar, Wild lupus, 22 anonymous edits

Foreign_exchange_market *Source*: http://en.wikipedia.org/w/index.php?title=Foreign_exchange_market *Contributors*: $1000000000ten0one1, (GD), 2010triennialsurvey, 2D, Abdul raja, Abdullais4u, Abj1, Acather96, Accesspig, Acmforex, Adamforex, Adamminstry, Adrie7, Ahunt, Alansohn, Alast0r, Alastairgbrown, Alberto.aff, Ale jrb, Alex771, AlexBraun, AlexiusHoratius, AllanManangan, Als7imy, Altasoul, Ammirajua, Andrew Maiman, Andyjsmith, Antandrus, Antifa82, Aotf01, Araignee, Arakesh999, Arthena, Ashlux, Asperal, Assarkareem, Aude, Auntof6, Barek, Barticus88, Barts1a, Bbatsell, Beedi, Beetstra, Ben Gaskin, Berlus, Bilalmanzoorrana, Bill37212, Billw2, Bisnisfx, Bjoleniacz, Blacksabbath4343, Blanchardb, BlazerKnight, Blue rose yy, Bluemoose, Bob walker99, Bobzchemist, Bonadea, Bongomatic, Borsaegypt, Boyd Reimer, Brandsen, Briantieling, Bruguiea, Buchandan, BusterBluth, Capricorn42, Casperdc, Catgut, Cerowyn, Chetannada, Chi0585, Chimidan, Chochopk, Chris the speller, Chris51659, ChrisEvans, Chrisch, Chrserad, ClamDip, Clerks, Clevercatskin, CliffC, Cliftonflack, Cneeds, Coopeajj, Coventgardenfx, Createmilk, Cst17, CurrencyMike, Cyun, DARTH SIDIOUS 2, DMacks, DOSGuy, DVdm, Da Vynci, Daddy32, Dancter, Daniel Renfoh, Danlaycock, Dargente, David Legrand, David Shay, DavidBoden, Davidjholden, Davidoff, Dawnseeker2000, Db099221, Deed89, Deepsnow, Delijaworld, Deon Steyn, DerHexer, Dfdferer22, DiamondJoeQuimby, Dijinkv, Dina Jones, Dinomite, Dionyziz, Discospinster, Djcyanide, Dneprolab, DocendoDiscimus, Doutrax, Dplayonline, DragonflySixtyseven, Dravecky, Drewwiki, Drmanotech, Dungodung, Dxantos, Dycedarg, Dyingsouls, Dylan Lake, E-mini, Easy101forex, Ecmoney, Ecocks1229, Edgar181, Editor334, Edward, Eforex, El C, Elockid, Elroch, EmadIV, Enivid, Epbr123, Erica j, Erik Lönnrot, Ersuzzi, Esejoker468, Espumadevidrio, Euryalus, EvanHarper, Excirial, Eyreland, Ezrakilty, F.h2010, FLarsen, FXWM, Fagallos, Fancitron, Feco, Feronnik, Figma, FinancialExpertise, Finnancier, Fireflyhu, Flamarande, Floshow25, Forex, Forex video, Forex-courses.us, ForexAficionado, ForexDude, ForexGuy, ForexReview, Forexbar, Forexmaster, Forexschool, Forextrader2011, Francs2000, Frankie1969, Fratrep, FreeKresge, Fwb44, Fx-daytrader, Fxdealer, Fxforex, Fxman, Fæ, G2010a, Gbsrd, Gene Nygaard, Ger123, Gilliam, Gld.signal, Globalwellnessteam, Gogo Dodo, GoingBatty, GraemeL, Gregdobbs, GregorB, Greswik, Grick, Ground Zero, GroveGuy, Grusaj, Gugustiuci, Gurch, H, Haca45, Hadal, Hamdouch03, Haus, Hit bull, win steak, Horatio, Htournyol, Hu12, Ianhowlett, Ifxekaterina, Igorn, Ilyasozgur, Imanizore, Immunize, Imnotminkus, Informedtrades, Innocent12345, Intelligentsium, Iridescent, Irineliul, Irishguy, Isatour, Island Monkey, J Di, J.delanoy, JForget, JLBernstein, Jackfitzgerald, Jackollie, Jackzhp, James Haughton, Japo, Jclfx, Jcurve100, Jenifan, Jennavecia, Jhaprashant, Jhone01kamil, Jibjibjib, Jim1138, JimS12, Jinex2012, Jmlk17, Jncraton, Jnrmedia, Joaquin777, Johan888, John0101ddd, JohnCD, Jojo16183, Jonath29, Joseph Solis in Australia, Josh Parris, Jredmond, Js2008, Jsxr7, Julesd, Jumpinginpuddles, Junglecash, Juni69, Jvcc, Jx-10, KMarie14, Kableash, Kallios, Karl Palmen, KathrynLybarger, Katieh5584, Kbh3rd, Kbrose, Kedi the tramp, Kidmercury, Kigali1, Kingjdub, Kingpin13, Kinitex, Kitsunegami, KoalaLovesWiki, Kog777, Kozuch, Kuru, Kwertii, LMB, Larklight, Laurenrach, Lavcom, Leafyplant, Lear's Fool, LeaveSleaves, LeeJohnOliver, Leewutzke, LegacyOfValor, Lemon-s, Leslie Mateus, Lesspaper, Lestatus, Leszek Jańczuk, Levineps, Lexxmarzain, Lgriot, Lightmouse, Lights, Limecastle, Linkspamremover, Linuzo, Liteforex.mx, LittleOldMe, Lmatt, Logan, Lotje, Lovealbatross, Luna Santin, Lupuskus, Lwcroslow, MER-C, MParaz, MRBigdeli, MaNeMeBasat, Madhero88, Mafiaforex, Magioladitis, MainFrame, Mainu30, Maksdo, Malayfx, Mamaoyot, Manikongo, Mark, MarquezComelab, Martianxo, Masken, Maslakovic, Materialscientist, Matt2727, Mattrix18, Mautinek, MaxEspinho, Mbahgat3000, Mbaluto, McSly, Mcarling, Mcduffodonnell01, Mdc-inmueb, Mdelfs, Meand, MegaSloth, Meisterkoch, Merikallos, Merlion444, Mgw, Michael Denmark, Michael Hardy, Mickey what a pity, Mike3608, Miltooon, MoForce, Mohamed Ouda, Monkeyman, Mormegil, MrArt, MrOllie, Mrmick73, Msankowski, Msmanikandan, Mwanner, NERIUM, NFH, Nakhoa, Nambe2, Natural Cut, Nealcardwell, Neelix, Neil916, NeilN, Neutrality, Nick Bencino, Nick Number, Nickhorder, Nicolebobbin, NikNaks, Ninja247, Nk, Norelaxation, Notgoogle, Notinasnaid, Novacatz, Nullnill, Nyse1982, Nyttend, OSX, Ohnoitsjamie, Old Guard, Oli Filth, OracleGD, Orangemike, OscarSebastian, Outlawcreations, OwenX, Oxymoron83, PBS27, PKT, PLA y Grande Covián, PTJoshua, Pagerank, Pandacomics, Papushin, PatrickFlaherty, Pearle, Pete8, Peter Tribe, Peter681, Pgk, Ph.eyes, Pharaoh of the Wizards, Phoenix-wiki, PiRSquared17, Pianojeff, Pill, Pilotguy, Pinkadelica, Pladook, Plain jack, Plastikspork, Pmhoey, Pmlineditor, Postdlf, Poweron, Pparazorback, Praddy06, Precisiontrader, Proofreader77, Pumpkin Pie, PurpleLime, Qarakesek, Qtia88, Quaeler, Quantumres, Quarl, Quentar, Quinjq, Quizimodo, Qwyrxian, RA0808, Rajrajmarley, Ramillav, Ray Van De Walker, Rdls01, Real Trade Group, Recreator, Redboylabs, Retired username, Rettetast, Rextech, Rgfolsom, Rich Farmbrough, Rixs, Rjwilmsi, Rl, Rnolds, Roeeric51, RonaldB, Ronhjones, Rpf 81, Rykerkim, SEO Swamee, SSZ, Sachindole, Samuk1000, Sango123, Sargdub, Sasuke Sarutobi, Scarian, Schzmo, Searchme, Seashorewiki, Sedimin, Segun1ng, Serbancea, Sergei Kazantsev, Sesu Prime, Shadow1, Shanec90, Shawncarpenter, Shay Stewart, Sherby33, Sheyne, Siggis, Simon12, Simon123, SkerHawx, Slawekb, Slyang, Smallbones, Smallman12q, Smileplz, Snesareva, SnowFire, Solipsist, SonicAD, South Bay, Spellcast, Spitfirebbmf, Spouima, Starfinances, StaticGull, Statoman71, Stefan2, Stefanvaduva, Steven Zhang, Stuffcool34, Suffusion of Yellow, Sumilord, SunCreator, Swatigoel9, THEN WHO WAS PHONE?, Tabletop, Tachyon01, Takeiteasyfellow, TaylorMarcella, Tbhotch, TejaraForex, Teles, Tellmemoreabout, Tennisnutt92, Terryfinotec, Texture, The Anome, The Thing That Should Not Be, The Transhumanist, Thecheesykid, Theluckyjerk, Themfromspace, Thingg, Tide rolls, Tiger888, Titoxd, Titus999, Tomas e, Tombomp, Tony Fox, Trade2tradewell, TraderFX, Tradingdude, Tradinghunter, Triangle eater, Trusilver, Ttmmblogger, Twidaleqwerty, UKFXONLINE, Ultramarine, Urdutext, Utcursch, V79, VPISoxFan, Vadimurazaev, Vailbrook, Vald, Varange2, Veinor, Versageek, Versus22, Vicsar, Virtuscience, Vishnu mohanan, Vision3001, Vivo78, Vklimko, Vpriest, Wai Wai, Wapman, Washburnmav, Wayne Riddock, Wbm1058, Wgibiz, White Lightning, Whufc48, WikHead, Wiki-winky, WikiDan61, Wikiacm, Wikidea, Wikireporter365, Wikixoox, Wild lupus, Wknight94, Woohookitty, X51xxx, Yahel Guhan, Yamaguchi, Yazdit1, Yosri, Youngreptile, Yourika, Yurik, Yuval613, Zahdanrino, Zikos750, Zondor, Zven, 1652 anonymous edits

Spot_contract *Source*: http://en.wikipedia.org/w/index.php?title=Spot_contract *Contributors*: 1862fun, 4realmoney, Alsocal, Anarchy, Angieangie, AtikuX, BKfi, Ben Vodden, Bezenek, Bfinn, Brad22dir, Cuberwhiz, DocendoDiscimus, Droplister, Elkman, Fancitron, Father McKenzie, Feco, Gardar Rurak, Garylhewitt, Gene Nygaard, Hadal, InsideGold, Julian Mendez, JustAGal, Kubanczyk, Lowellian, Mark Arsten, Mattis, Mulad, Mvpolis, Patstuart, Philip Trueman, Psb777, Rd232, Rigadoun, Rpf 81, Sargdub, Seksen iki yüz kırk beş, Shoefly, SimonP, Smallbones, U608854, Yurik, Z10x, Zzuuzz, 25 anonymous edits

Electronic_trading *Source*: http://en.wikipedia.org/w/index.php?title=Electronic_trading *Contributors*: Arthur Frayn, AssiToro, Astellix, Asterix.wins, AvicAWB, Beland, Bmathis, Brian0918, Carly McCormick, Chandrasekhart, Criffin, DMacks, Deville, Edward, Gary-Pav, Gco, GoingBatty, HotBridge, Hroðulf, Hu12, II MusLiM HyBRiD II, Jeff G., John of Reading, Joseph Solis in Australia, Jpbowen, Kearney09, Koavf, Lmatt, Mandarax, MontyPh, Nave.notnilc, Netrapt, OccamzRazor, Openstrings, Pascal.Tesson, PaulnelsonIII, Peoplesunionpro, Per Ardua, PhilKnight, Professorpkjain, Quintote, R'n'B, Rayngwf, Sargdub, Smallbones, Snowfalcon, Vecter, Wikipelli, Willisja, 66 anonymous edits

Software *Source*: http://en.wikipedia.org/w/index.php?title=Software *Contributors*: 021-adilk, 12dstring, 16@r, 194.237.150.xxx, A Stop at Willoughby, A.K.R., ABC1A1, ABF, ALargeElk, AManWithNoPlan, Aardvarkzz, Aatayyab, Abby, Abufaisal65, Achowat, Acroterion, Addshore, Ademsaykin, AdjustShift, Adriancjr, Ah2190, Ahadrt, Ahoerstemeier, AlMac, Alan012, Alansohn, Alasdair, Alazlogexi, Aleemqureshi, Ali K, AlisonW, AlistairMcMillan, Allen4names, Allens, Allmightyduck, Alphachimp, Amalhotra124, Anabus, Andre Engels, Andrea105, Andreas Kaufmann, Andrejj, Andrewspencer, Angela, Angela 2502, Anikingos, Ankur 360, Antandrus, Aoratos, Ap, Apparition11, Aqeelbilal, ArchonMagnus, ArglebargleIV, Arkrishna, Arman Cagle, Arnabatcal, Artichoker, Ashawley, Asiananimal, AuburnPilot, AussieLegend, Autocratique, Avono, Aweiredguy, Axcelis555, AznBurger, BD2412, Backslash Forwardslash, Bact, BalderV, Barek, Bean2thousand, Bearly541, Ben-Zin, Betterusername, Bevo, Big Bird, Big milsy, BigDunc, Binary TSO, Blarrrgy, Bobo192, Bonadea, Bongwarrior, Booyabazooka, Boriszex, Boxplot, Brat32, Brent Gulanowski, Brinnington, Brucevdk, Brusegadi, Bryan Derksen, Bubba73, Burnisk, Butros, CIreland, CWenger, Calcifero, Caliboobool, Can't sleep, clown will eat me, CanadianLinuxUser, Canaima, Canterbury Tail, Cargils02, Carl007, Casperdog2227, Catgut, Causa sui, Cburress, Centrx, Chealer, Christian List, Christian75, Chun-hian, Chuunen Baka, Chzz, Citypanther, Cloudyed, Clovis Sangrail, CommonsDelinker, Conan, Conversion script, CoolD, Coolcaesar, Cpl Syx, Crazykr, Crossmr, Cst17, Cunya, Cybercobra, Cyberstrike3000X, Cyrius, DARTH SIDIOUS 2, DBHunter, DJ Clayworth, DLMahnken, DSP-user, DURWHAT, DVdm, Damian Yerrick, Dan Fuhry, Dan100, Danakil, Daniel5127, DanielEng, Darkevilfairy, Darrendeng, Darth Panda, Dave6, Davennmarr, Daveydweeb, DavidWBrooks, Davie4125, Dayewalker, Dbstommy, DeadEyeArrow, Deggalega, Delirium, Denny, DerHexer, Derek farn, Dethme0w, Dffgd, Dibcom, Dina, Dinhtuydzao, Discospinster, Dispenser, DivineAlpha, Dj stone, DmitTrix, Dominator09, Doriftu, Dougofborg, Drawnman247, Dreamatalana, Drewster1829, Drmies, Dugmn, Dwayne, Dycedarg, Dynix, E. Sn0 =31337=, E2eamon, EdBever, Edcolins, Edderso, Edivorce, Edlin, Eeekster, Egbsystem, Ehheh, Eidetic Man, ElBenevolente, Elassint, Elizium23, Ellenaz, Emana, Emvn, Entropy, Epbr123, Equendil, Erdal Ronahi, EricWesBrown, Esoteric Rogue, Ethine, Everyking, Ex-dot, Excirial, Ezubaric, FJPB, Fang Aili, Fastily, Favonian, Fazilati, Fctk, Feedmecereal, Felyza, Fenice, Fern80, Ff1959, Flarn2006, Flash man999, Flipjargendy, Flubeca, FlyingToaster, Frap, Fratrep, Fred Bradstadt, Fredrik, Fredtheflyingfrog, Freeeeeesoft, Friday, Frip1000, Frosted14, Fubar Obfusco, Funnysens, Fyyer, GDallimore, GSCC, GTBacchus, Gabecuevas, Gadfium, Gaius Cornelius, Galoubet, Gardar Rurak, Gary King, Gazpacho, Geneb1955, Georgia guy, Gfoley4, Giftlite, GilbertoSilvaFan, Gilliam, Ginsengbomb, Gogo Dodo, Gordmoo, Grafen, Greenrd, GregorB, Gronky, Gtg204y, Guanaco, Gueldenberg, Guppy, Gurch, Gwen-chan, Gwernol, Hackguard, Hadal, Hamiltøn, Hans Dunkelberg, Happyisenough, Harry, Harvester, Hauberk, Haunting The Better, Hayabusa future, Hectorthebat, Hello32020, Hellosandimas, HenryLi, Hmains, Hmcnally, Hno3, Howdoesthiswo, Howling wolf of the jungle, Hu12, Hutcher, Hydrogen Iodide, Icairns, Iffikhaarrr, Iknowyourider, Imnotminkus, Imroy, IngerAlHaosului, Instinct, Intelligentsium, Into The Fray, Iridescent, Ishara665g, Itsmine, Ixfd64, J Di, J. Spencer, J.delanoy, JLaTondre, JMetzler, Jaan513, Jackaranga, Jacob.jose, Jacqueline7894y, Jam2k, Jamesiemiller, Jan1nad, January, Japonca, Jarda-wien, Jasper Deng, Jay, Jcrwaford5, Jcuadros, Jcw69, Jeff G., Jehochman, Jennapresley, Jeremy Visser, JeremyA, JesseW, Jester7777, Jh51681, Jiang, Jiddisch, Jim1138, Jiy, Jmabel, JoanneB, Johan1298, JohnCD, JonHarder, Jonnabuz, Jop, JordoCo, Joy, Jpbowen, Jpgordon, Jreferee, Jsheadixon, Jtalledo, Jtbatalla, Jusdafax, Jusjih, Justin Eiler, Jóna Þórunn, K.Nevelsteen, K.lee, KFP, KGasso, Kajasudhakarababu, Karol Langner, Kathmandu2007, Kc03, Keilana, Kelldall, Kelly Martin, Kenny sh, Kenyon, KevinCuddeback, Kingpin13, Kinneyboy90, Klilidiplomus, Knodir, Knownot, Knucmo2, Koavf, Kornxi, Kostisl, Kozuch, KrakatoaKatie, Krawi, Ktanzer, Kuldeep06march, Kumioko (renamed), Kurowoofwoof111, Kuru, L Kensington, La Pianista, La comadreja, Labscientist, Lagalag, LaggedOnUser, Lajpatdhingra, Landroo, Lareina3656y, LauriO, Leandrod, LeoNomis, Liao, Liberty Miller, Liftarn, Ligulem, LilHelpa, LinDrug, Lindberg G Williams Jr, Listlist, LittleOldMe, Livajo, Lookingchris, Lucianob, Luk, Luna Santin, Lunaumbrax, Luntrasul, Lupo, M.O.X, M4gnum0n, MC10, MD66, MER-C, Macy, MaggieMidnight, Majilis, Majorbrainy, Malcolm Farmer, Malcolma, Mandarax, Manik762007, Manionc, Manishgoyal.indian, Maralia, Marcelnache, Marek69, MarioS, Markaci, Markwaugh117, Martarius, Master&Expert, Matt Britt, Matt j fox, Matthias M., Matusz, Mav, Maximaximax, Maximus Rex, Mayur, Mcdonald.ross5, Mdd, Mdebets, MeekSaffron, Meelar, Mendalus, Mentifisto, Mermaid from the Baltic Sea, Message From Xenu, Mgrand, Michael Drüing, Michael Hardy, Michal Jurosz, Miguel.mateo, Milan Keršláger, Mild Bill Hiccup, Mindmatrix, Mindspillage, Minesweeper, Minghong, Mini-Geek, Miquonranger03, MithrandirAgain, Mjchonoles, Mkdw, Monobi, Mpete510, Mr Barndoor, MrDKing, MrDolomite, MrOllie, Muffhen, Murray Langton, Mwanner, Mwilso24, Mxn, Myanw, N n abc123, Nabeth, Naik5abhi, Nanshu, Narayana vvss, Nascar1996, Nathanielrichards, Navstar, NawlinWiki, Nbarth, Neil916, Neurophyre, Nharipra, Nick Number, Nickels360, Nikai, Nike8, Nima1024, NinjaCross, Nixdorf, Nixeagle, Nk, Nmrd, Normxxx, Northamerica1000,

Nosferatus2007, Nourybouraqadi, Nsaa, NuclearWarfare, Nunquam Dormio, NurAzije, Nurg, Ocaasi, Oda Mari, Ohnoitsjamie, Oicumayberight, Old Moonraker, Oldwes, Oleg Alexandrov, Oliver Lineham, Omicronpersei8, One more night, OrgasGirl, Oxymoron83, P.deshmukh09, P99am, PEH2, PIrish, Page Up, Paranoid, Park3r, Party, Passargea, Paul August, Paul Foxworthy, Paul Niquette, Pcap, PeaceAnywhere, Perkinsleslie, Peruvianllama, Peter Winnberg, Petrb, Pgk, Pgr94, Phatom87, PhilHibbs, PhilKnight, Philip Howard, Philip Trueman, Phillip Ca, Piano non troppo, Piasa1, Pigduckmeatshroom, Pilotguy, Pip2andahalf, Pnm, Poccil, Pohatu771, Pointillist, Poo1000, Pooryorick, Prahim, Priyadarshi.pratyush, Priyaranjanchampatiray, Programmer13, Prohlep, Psantora, Public Menace, Puchiko, Pufferfish101, Puneet1507, Purgatory Fubar, Pyfan, Quinobi, Quinsareth, QuintusQuill, R. S. Shaw, RHaworth, Ragha joshi, Ragib, Rainier3, RainierHa, Rajeshmagic, Random89, Raul654, Ray Van De Walker, Raylu, Rbreen, Reaper Eternal, Rebroad, Recnilgiarc, Recognizance, RedWolf, Redmercury82, Reedy, Rehnn83, Reisio, Remember the dot, Rettetast, RevRagnarok, Revised, Rgill, Rholton, Rhye123, Riana, Rich Farmbrough, Richard cocks, RickK, Rihannoufal, Ripogenus77, Rl, Rodhullandemu, Rogger.Ferguson, Roland2, Ronhjones, RossPatterson, Roybb95, Rurik, Rwwww, Ryulong, S.K., S0aasdf2sf, SC979, SF007, ST47, SaadTeenager, Sadeq, Sadi Carnot, Safalra, Safarj, Saleem110, Salmank120, Salsa Shark, Sango123, SchfiftyThree, SchreyP, Scientus, Scrool, Seaphoto, Seb az86556, Seidenstud, Senator Palpatine, Setveen, Seven.cardwell, SewerCat, Shadowjams, Shakya ind, Shaluhijas, Siggy28, Sigma 7, Sigondronggondrong, Siliconov, Silivrenion, SimonP, Sintonak.X, Sitarherophil, Sivaguru411, Sjö, Skarebo, Skizzik, Skybon, Slingerjansen, Slowking Man, Slowmo1993, Smack, Smalljim, Snowmanradio, Sokrato, Soler97, Some jerk on the Internet, Sorryranga94, South Bay, Speaksleft, Spectrogram, Spoon!, SpuriousQ, Standalone0109, Starkiller88, Stephenb, SteveWringsfield, Steven Zhang, Stewartadcock, StuffOfInterest, SudoGhost, Supadawg, Super-Magician, Supersteve04038, Suruena, SusanLesch, SwirlBoy39, Syrthiss, T, T3h 1337 b0y, T4tarzan, TC Cannon, THEN WHO WAS PHONE?, TPK, TaintedMustard, Tangent747, Tasja, TastyPoutine, Tbhotch, Tejas81, Teles, Teryan2006, Th1rt3en, Tharcore, The Anome, The Earwig, The High Fin Sperm Whale, The Rambling Man, The Thing That Should Not Be, The Transhumanist, The Transhumanist (AWB), TheBiaatch, Thelb4, Thingg, This acccount is 4 vandalism, Thumperward, Thurak13, Tide rolls, Tiffu, Tigershrike, Tim Q. Wells, Timhowardriley, Timothy Neilen, Tommy2010, Topbanana, Torc2, Torinor, Traroth, Trevor MacInnis, Triona, Triwbe, True Genius, Trueshow111, Trusilver, Turlo Lomon, Tuxide, Twsx, Uncle Milty, Unyoyega, Uriyan, Utuado, Vale Len, Vbigdeli, Velella, Venkisree, Versageek, Versus22, Vibhijain, Virtualenv, Virtualsfera, Vishwajith33, Viskonsas, Vrenator, W Nowicki, WLU, Wavelength, Wayne Slam, Wayward, Wbm1058, Web20image, Werdna, Weregerbil, Wernher, Wexhammer, Weyes, Whitew123, Wiki alf, Wikicrazier2011, Wikidudeman, Wikieditor06, Wikijens, Wimt, Winterst, Wireless friend, Wm, WojPob, Woohookitty, WookieInHeat, Wshun, Wysprgr2005, XP1, Xdenizen, Xevious, Xionbox, Xplodercop, Yamamoto Ichiro, Yasirilyas, Yidisheryid, Yoooder, YountLor, Yworo, Zarkos, Zatya34, Zedlik, ZeroOne, Zerokewl, Zhou Yu, ZimZalaBim, Zybez, Александър, Евгени Симеонов, Петър Петров, ТимофейЛееСуда, 1859 anonymous edits

Electronic_trading_platform *Source*: http://en.wikipedia.org/w/index.php?title=Electronic_trading_platform *Contributors*: Bearcat, Cretog8, Empty Buffer, Fabrictramp, Firetinder, Garion96, Hashem sfarim, Ivliev, JRaeside, JamesBWatson, John of Reading, Koppas, LKP Securities, Lmatt, Macaddct1984, Malcolma, Nagle, Neelix, Ohnoitsjamie, PamD, Penqem, Plidsyu, R'n'B, Sargdub, Schuym1, Tedgregory, The Interior, Theroadislong, Tradechat3, Woohookitty, 21 anonymous edits

CMC_Markets *Source*: http://en.wikipedia.org/w/index.php?title=CMC_Markets *Contributors*: Blanchardb, CMC Markets PLC, Chzz, DMCer, East718, Edgar181, Edward, Fabrictramp, Gsterz, Hydrargyrum, Jamirocrazy, Jgombos, John of Reading, Khazar2, Klf uk, Kmbharadia, LilHelpa, NotaDestination, Phil Bridger, Porter2010, SFC9394, Sargdub, Sfan00 IMG, Sspyrou, Trident13, Vald, Wikimanchild00, Yourselfe, 27 anonymous edits

Portfolio_manager *Source*: http://en.wikipedia.org/w/index.php?title=Portfolio_manager *Contributors*: Cander0000, Chzz, DanielRigal, Edward, Evans1982, Pah799, Shawnc, Sliderule, WikiaPage, Woohookitty, 14 anonymous edits

Saxo_Bank *Source*: http://en.wikipedia.org/w/index.php?title=Saxo_Bank *Contributors*: Aelfthrytha, Aspects, Barticus88, Ben Ben, Boing! said Zebedee, Burpelson AFB, Cameron Scott, Christianvinter, Cielbleu, Crashwhiled2, CygnusPius, Details.ch, Dozo, Drdarst, Drewwiki, Dsp13, Edward, ErrantX, Evadanit, Fair and unbiased, Favonian, Forextrue, Fratrep, GigaVL, GoingBatty, Hebrides, Hu12, Hydrargyrum, Jjozolins, Josh Parris, Kaldari, Katharineamy, Kirstine Andersen, Kirstine11, Kolporti, Lamro, Lavcom, LilHelpa, Magioladitis, Mardus, Mitsaxo, Najskapati, Npa1922, Nurg, Ohconfucius, Onorem, Orange32, Ph.eyes, PhilipHeyman, Pontoppidan, PrimeHunter, Racklever, Ramblersen, Rich Farmbrough, Rik63, Rjwilmsi, Sargdub, Sasper, Saxofraud, Saxoinside, Skarebo, Sledge 1981, Stifle, Tabletop, Tedder, TicketMan, Tim1357, Topbanana, Uae expat, UltraMagnus, Vitassa, Welsh, Woohookitty, Xezbeth, Yoenit, 105 anonymous edits

Matchbook_FX *Source*: http://en.wikipedia.org/w/index.php?title=Matchbook_FX *Contributors*: BD2412, Bongomatic, Chris the speller, CommonsDelinker, Diannaa, Eforex, ForexAficionado, GodfatherOfFX, Malcolma, Mild Bill Hiccup, Ph.eyes, Sargdub, 8 anonymous edits

MetaTrader_4 *Source*: http://en.wikipedia.org/w/index.php?title=MetaTrader_4 *Contributors*: Antarctic-adventurer, Asdiop1000, Belovedfreak, Ben Ben, Bswen1, Bunnyhop11, Czarkoff, Download, EATrader, Edward, Enivid, Ether13, I dream of horses, Jcaraballo, Justin W Smith, Karotic, Lmatt, Logan, Mild Bill Hiccup, Molanis, Mortense, Olimpiu stefan, Onliner 1010, Rich Farmbrough, Sargdub, Sharkface217, Smallbones, Texasshaun, Tschonen, Vadimurazaev, Zellfaze, Пrate, 23 anonymous edits

MetaQuotes_Software *Source*: http://en.wikipedia.org/w/index.php?title=MetaQuotes_Software *Contributors*: Argey, Czarkoff, Download, Enivid, Eumolpo, Jamietw, Jim1138, Jørdan, Kuru, Michaelpillos, Muhandes, Olimpiu stefan, QuentinUK, Renat.Shafigullin, Sargdub, 16 anonymous edits

Image Sources, Licenses and Contributors

Image:G foreign exchange market turnover.gif *Source*: http://en.wikipedia.org/w/index.php?title=File:G_foreign_exchange_market_turnover.gif *License*: unknown *Contributors*: Notgoogle

File:Flag of Germany.svg *Source*: http://en.wikipedia.org/w/index.php?title=File:Flag_of_Germany.svg *License*: unknown *Contributors*: Anomie

File:Flag of the United Kingdom.svg *Source*: http://en.wikipedia.org/w/index.php?title=File:Flag_of_the_United_Kingdom.svg *License*: unknown *Contributors*: Anomie, Mifter

File:Flag of Switzerland.svg *Source*: http://en.wikipedia.org/w/index.php?title=File:Flag_of_Switzerland.svg *License*: unknown *Contributors*: User:-xfi-, User:Marc Mongenet, User:Zscout370

File:Flag of the United States.svg *Source*: http://en.wikipedia.org/w/index.php?title=File:Flag_of_the_United_States.svg *License*: unknown *Contributors*: Anomie

File:Flag of Europe.svg *Source*: http://en.wikipedia.org/w/index.php?title=File:Flag_of_Europe.svg *License*: unknown *Contributors*: User:-xfi-, User:Dbenbenn, User:Funakoshi, User:Jeltz, User:Nightstallion, User:Paddu, User:Verdy p, User:Zscout370

File:Flag of Japan.svg *Source*: http://en.wikipedia.org/w/index.php?title=File:Flag_of_Japan.svg *License*: unknown *Contributors*: Anomie

File:Flag of Australia.svg *Source*: http://en.wikipedia.org/w/index.php?title=File:Flag_of_Australia.svg *License*: unknown *Contributors*: Anomie, Mifter

File:Flag of Canada.svg *Source*: http://en.wikipedia.org/w/index.php?title=File:Flag_of_Canada.svg *License*: unknown *Contributors*: Anomie

File:Flag of Hong Kong.svg *Source*: http://en.wikipedia.org/w/index.php?title=File:Flag_of_Hong_Kong.svg *License*: unknown *Contributors*: Designed by

File:Flag of Sweden.svg *Source*: http://en.wikipedia.org/w/index.php?title=File:Flag_of_Sweden.svg *License*: unknown *Contributors*: Anomie

File:Flag of New Zealand.svg *Source*: http://en.wikipedia.org/w/index.php?title=File:Flag_of_New_Zealand.svg *License*: unknown *Contributors*: Achim1999, Adabow, Adambro, Arria Belli, Avenue, Bawolff, Bjankuloski06en, ButterStick, Denelson83, Donk, Duduziq, EugeneZelenko, Fred J, Fry1989, Hugh Jass, Ibagli, Jusjih, Klemen Kocjancic, Mamndassan, Mattes, Nightstallion, O, Peeperman, Poromiami, Reisio, Rfc1394, Shizhao, Tabasco, Transparent Blue, Väsk, Xufanc, Zscout370, 35 anonymous edits

File:Flag of South Korea.svg *Source*: http://en.wikipedia.org/w/index.php?title=File:Flag_of_South_Korea.svg *License*: unknown *Contributors*: Various

File:Flag of Singapore.svg *Source*: http://en.wikipedia.org/w/index.php?title=File:Flag_of_Singapore.svg *License*: unknown *Contributors*: Various

File:Flag of Norway.svg *Source*: http://en.wikipedia.org/w/index.php?title=File:Flag_of_Norway.svg *License*: unknown *Contributors*: User:Dbenbenn

File:Flag of Mexico.svg *Source*: http://en.wikipedia.org/w/index.php?title=File:Flag_of_Mexico.svg *License*: unknown *Contributors*: User:AlexCovarrubias

File:Flag of India.svg *Source*: http://en.wikipedia.org/w/index.php?title=File:Flag_of_India.svg *License*: unknown *Contributors*: Anomie, Mifter

Image:Indian Rupee symbol.svg *Source*: http://en.wikipedia.org/w/index.php?title=File:Indian_Rupee_symbol.svg *License*: unknown *Contributors*: User:Orionist

File:Equities usd.JPG *Source*: http://en.wikipedia.org/w/index.php?title=File:Equities_usd.JPG *License*: unknown *Contributors*: TheGeekKnows.com

Image:Operating system placement.svg *Source*: http://en.wikipedia.org/w/index.php?title=File:Operating_system_placement.svg *License*: unknown *Contributors*: Photos:GolfthemanGolftheman

File:Deutsche-boerse-parkett-ffm008.jpg *Source*: http://en.wikipedia.org/w/index.php?title=File:Deutsche-boerse-parkett-ffm008.jpg *License*: unknown *Contributors*: User:Dontworry

File:CMC LOGO rgb 123 70.png *Source*: http://en.wikipedia.org/w/index.php?title=File:CMC_LOGO_rgb_123_70.png *License*: unknown *Contributors*: NotaDestination, Sfan00 IMG, Skier Dude, Ww2censor

File:Increase2.svg *Source*: http://en.wikipedia.org/w/index.php?title=File:Increase2.svg *License*: unknown *Contributors*: Sarang

File:Decrease2.svg *Source*: http://en.wikipedia.org/w/index.php?title=File:Decrease2.svg *License*: unknown *Contributors*: User:Sarang

Image:SaxoBankLogo.gif *Source*: http://en.wikipedia.org/w/index.php?title=File:SaxoBankLogo.gif *License*: unknown *Contributors*: Original uploader was Dozo at en.wikipedia

File:Matchbook-FX logo.gif *Source*: http://en.wikipedia.org/w/index.php?title=File:Matchbook-FX_logo.gif *License*: unknown *Contributors*: User:ForexAficionado

File:MetaTraderPlatform.gif *Source*: http://en.wikipedia.org/w/index.php?title=File:MetaTraderPlatform.gif *License*: unknown *Contributors*: Sargdub

File:Fibretracement.png *Source*: http://en.wikipedia.org/w/index.php?title=File:Fibretracement.png *License*: unknown *Contributors*: MetaQuotes Software Corp.

File:MetaQuotes Logo.png *Source*: http://en.wikipedia.org/w/index.php?title=File:MetaQuotes_Logo.png *License*: unknown *Contributors*: Sargdub, We hope

MIX
Papier aus verantwortungsvollen Quellen
Paper from responsible sources
FSC® C105338

Printed by Books on Demand GmbH, Norderstedt / Germany